Microtunneling and Horizontal Drilling

Microtunneling and Horizontal Drilling

French National Project "Microtunnels"

Recommendations

FSTT

French Society for Trenchless Technology

affiliated society of
ISTT
International Society for Trenchless Technology

ISTE

First published in Great Britain in 2004 by Hermes Science Publishing Ltd
Published with revisions in Great Britain and the United States in 2006 by ISTE Ltd

ISTE Ltd
6 Fitzroy Square
London W1T 5DX
UK
www.iste.co.uk

ISTE USA
4308 Patrice Road
Newport Beach, CA 92663
USA

First South Asian Edition 2007

Library of Congress Cataloging-in-Publication Data

Cemito franais des travanx sans frenchee
Microtunneling and horizontal drilling; French national project "microtunnels" guidelines/ FSTT.
p. cm.
ISBN 13: 978-1-905209-00-2
1. Trenchless construction. 2. Tunneling. I. Title.
TA815.C66 2006
624.I'93—dc22

2005033972

British Library Cataloguing-in-Publication Data
A CIP record for this book is available from the British Library
ISBN 10: 1-905209-00-2
ISBN 13: 978-1-905209-00-2

Printed in Brijbasi Art Press Ltd., I-72, Sector-9, Noida, U.P. India.

Table of Contents

The research and study programme was financed by the members of the National Project: ANTEA – AQUAREX – BONNA – BORIE SAE – CAMPENON BERNARD – CERIB – CGG – CHANTIERS MODERNES – CONDAT – DEVIN LEMARCHAND – the NANCY District – EDG – FOUGEROLLE/BALLOT – Gas of France – GEOMEGA – GEOSCAN – INSA of LYON Geotechnical laboratory – INSTITUT FRANÇAIS DU PETROLE – JF TECH – NANCY Geomechanics laboratory – LCPC – QUILLERY – RATP – SADE – SCETAUROUTE – SPIE CITRA – TERRASOL – UNION TRAVAUX – University of LILLE/GERFEC – PARIS VI University/Geophysics Department – VALENTIN, with the support of DRAST.

The National Project included an executive committee presided over by Mr. Mermet. Technical supervision was ensured initially by Jean-Pierre Henry and then by Alain Guilloux since 1996.

These recommendations have been prepared by a drafting committee consisting of:

- Djamel Ait Aissa (SIARCE),
- Sophie Areia (SNCF),
- Michel Audouin (FSTT),
- Anne-Lise Beaucour (IUP de Cergy-Pontoise),
- Jean-Pierre Brazzini (GDF),
- Frédéric Bultel/Richard Tuphe (SCETAUROUTE),
- Jack Butterworth (LMR Drilling),
- Dominique Commery (Tracto Techniques),
- Stéphane Delafontaine (Radiodétection),
- Philippe Delorme (GDF),
- Damien Deppner (REHAU),
- Michel Guérin (Société française des bentonites, SFDB),
- Alain Guilloux (Terrasol),
- Richard Kastner (INSA de Lyon),
- Jacques Lacombe (SADE),
- Michel Lamy (retraité REHAU),
- Christian Legaz (DDE du Val-de-Marne),
- Eric Lessault (SADE),
- Frédéric Ouvry puis Jean Piraud (ANTEA),
- Anne Pantet (ESIP, Ecole supérieure des Ing. de Poitiers),
- Daniel Philippe (SADE),
- Patrice Schneider (Cogeprec),

– Bernard Sustrac (BCM),

– Michel Vincent (Forage 21),

– Roger Wilkinson (Wise),

based on 31 technical reports and 26 status reports of the National Project (see bibliography).

PART I

Microtunneling

Chapter 1

Introduction to Guidelines: Subject and Fields of Application

1.1. General introduction of "trenchless technology"

These guidelines apply to the construction of structures by microtunneling, which is a part of the "trenchless technologies". These techniques are currently used in urban areas in an age where environmental degradation has become an ever-increasing concern. It involves creating new networks or repairing existing ones (water, sanitary drainage, electricity, gas, etc.) by minimizing the impact on surface sites. This reduces the inconvenience caused to the users by "open trench" (or "cut and cover") work, which requires an excavation along the full length of the area worked on.

Even though, most often, it does not involve work to the same extent as that for large sites such as the underground or motorways, its importance in terms of linear structures entirely justifies our interest in it, as much for its economic impacts as for its close overlapping with social life.

It is necessary first to specify a definition which helps better determine the field of application of this work. Of course, the term "trenchless" is the opposite of "open trench" work, but it is also used for the installation of networks of small diameter, which are called "inaccessible", particularly where a worker cannot get into the networks in normal working conditions: it is generally accepted that the upper limit is approximately 1,200 mm in diameter. We are interested in underground structures where the construction requires remote controlled techniques because the site can neither be accessed from the surface ("trenchless") nor accessed from the inside (inaccessible).

It is common in the field of "trenchless digging" to distinguish between various procedures, for which the techniques used are very different and whose fields of application are equally diverse. Firstly, new construction projects and old renovation projects have to be distinguished.

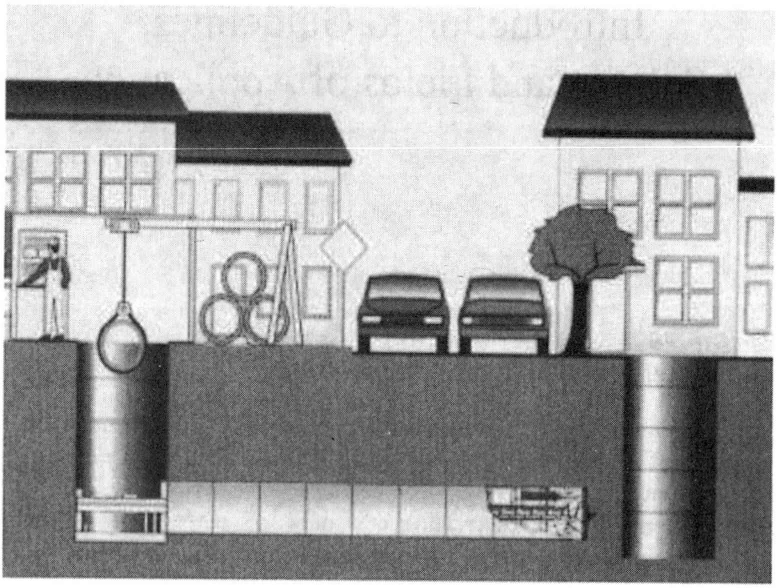

Figure 1.1. *Diagram of a microtunneling site*

a) The new structures involve the creation of networks where nothing exists and again for this, two categories can be considered corresponding to very different techniques:

– *microtunneling* (see Figure 1.1) is used for networks with diameters generally ranging from 500 to 1,500 mm and which can go up to 2,000 mm. The boring machines resemble Tunnel Boring Machines (TBMs) of large diameters, and have the special feature of being miniaturized and remote controlled, which means that they can be operated without any human intervention inside the machine. The machines operate along a linear trajectory at variable depths ranging from just a few meters to more than ten meters and along a length of approximately 100 to 150 m: thus, they have to be installed through shafts dug from the surface up to the depth of the project. This enables the machines and its pipes to be sunk to the depth required for the project and then be recovered at the outlet.

– *horizontal drilling* is used in general for urban networks of small diameter (100 to 500 mm) as well as for pipelines of up to 1,000 mm in diameter. The technique is derived from traditional drilling with the added ability to locate the position of the drilling head in the plane and/or in depth and above all to correct the direction if there is a major deviation from the trajectory. It mostly relates to low depth networks (a few meters at the most) but can, in some cases with appropriate equipment, be used for installing pipes at greater depths. This is not covered in this discussion.

b) The renovation of old structures is used for existing networks whose ageing condition does not permit them to properly fulfill the functions they were intended for. Thus, it is necessary to put them into normal operating conditions with one of the following techniques:

– replacement, by creating a parallel new network (this brings us to the previous cases),

– renovation, by restoring damaged pipes over large curbsides,

– repair, by selective restoration.

Many different techniques that are not mentioned in the current Guidelines may still be distinguished.

1.2. History and characteristics of microtunneling methods

The microtunneling techniques are relatively recent: the first boring machines were used in Japan during the 1970s. In France, the first site was constructed in 1989 in the Val-de-Marne department at the instigation of the Water and Sanitary Drainage Services (Mermet *et al.,* 1991). Currently, the development of this technique varies greatly from country to country: in Japan the curbside reaches several hundred kilometers per year; in Germany and the UK it spans several dozen kilometers whereas in France it is less than 10 km.

Before describing the microtunneling techniques in greater detail, it is important to state that their implementation requires a change in "culture" on the part of various contributors. In fact, if the installation techniques with trenches result in general from traditional methods which are mostly of relatively low technical nature, it should be kept in mind that the trenchless techniques more closely resemble the methods of underground work in the broad sense and therefore require a highly technical approach.

Amongst the characteristics of underground work which form part of microtunneling, we will list the following main elements:

– the equipment is relatively sophisticated; for this reason their implementation requires a good knowledge of their functioning and the maintenance aspects are very important as well,

– their optimal functioning depends greatly on the suitability of the choice of various components of the machine, the nature and the performance of the ground to be crossed and the ability of the operator to adapt to the local conditions,

– for this reason, prior knowledge of the ground to be excavated is essential for the success of the project: geotechnical investigations thus become an important element in the project design,

– finally, the small diameter of structures and the low depths at which they have to be set-up, in embankments or geological formations on the surface, make the digging particularly sensitive to numerous natural (blocks) or artificial (old foundations, existing structures) heterogeneities. The investigation methods should therefore be able to detect these heterogeneities.

1.3. Purpose of the guidelines

These different preambles are obviously not designed to threaten the design technicians and decision-makers so that they are forced to do away with the trenchless techniques *a priori*, but rather to make them aware of the minimum precautions to be taken when initiating such projects. The purpose of these guidelines is to give the various parties sufficient knowledge and the necessary elements for the success of the projects.

They are aimed at assisting the following:

– contracting authorities (owners) that wish to know the potential of these techniques,

– engineers who have to design the projects,

– design offices, particularly geotechnological design, that need to recognize such projects,

– companies who generally know the techniques well but who may need some "reference material",

– finally, the manufacturers of the pipes concerned about supplying the equipment most suited to the tool and the method used.

We must emphasize that the trenchless digging techniques in France were the subject, during the 1990s, of "national research projects" involving owners, engineers, specialized companies, design offices and research laboratories with partial government funding, so as to better understand the performance of structures and optimize the projects. It is in particular the French National Research Project

"Microtunnels" that includes microtunneling work and horizontal drilling of which the current Guidelines are an outcome. This National Project developed various theoretical as well as experimental researches from the monitoring of various sites; the results of this research are available in 60 technical and comprehensive reports, which are listed in the bibliography.

The publications on this subject are still relatively rare, especially global publications. The book by Stein *et al.* (1989) was the first that provided a comprehensive study of boring machines as well as the NO-DIG conference acts, which have been held annually since 1989.

Finally the existence of FSTT[1] (*French Society for Trenchless Technology*), an association that groups together the main players in the field of trenchless digging, has been promoting this technology since 1990. The FSTT has undertaken the management of the French National Research Project "Microtunnels" and the reader interested in finding out more can obtain a vast documentation on this subject.

1. FSTT: 4 rue des Beaumonts – F 94120 Fontenay-sous-Bois. Phone: (00-33) 1 53 99 90 20 – Fax: (00-33) 1 53 99 90 29 – email : fstt.paris@wanadoo.fr – website : www.fstt.org.

Chapter 2

Techniques and Theory of Operation for the Installation of Pipes by Microtunneling

2.1. General information

The microtunneling technique enables the installation of pipelines in rectilinear sections of lengths ranging generally from about ten meters to more than a hundred meters. Making it possible to follow the direction and slope, this procedure is particularly suitable for wastewater networks with gravitational flow and is also used for other networks such as drinking water and telecommunications.

The principle of microtunneling (see Figure 2.1) is similar to that of TBMs, whose technique is suitable for inaccessible pipelines of diameters ranging from 400 mm to 1200 mm, which now extend to more than 2000 mm. Like the TBMs, boring machines have a shield that ensures the temporary support of the excavation site, a rotary excavator fitted with cutting tools and a mucking system enabling the application of a confining pressure on the face.

As the final lining cannot be done under the shield for reasons of obstruction, it is made up of pipes driven one after the other from the starting shaft. It is this set of pipes, preceded by the boring machine, that are driven into the ground with the help of a thrust frame located in the start shaft. The operator operates the various systems of the machine from the surface. The trajectory, which is rectilinear from the starting shaft (consisting of the pushing frame and its safety pillar) to the exit shaft, is followed by pointing a laser beam on a target on-board the boring machine. The operator can correct the deviations in the trajectory by modifying the direction of the machine head.

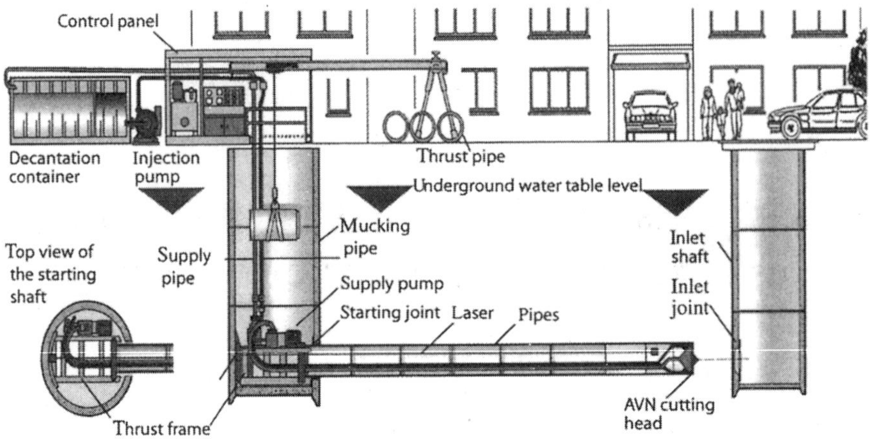

Figure 2.1. *Microtunneling principle (Herrenknecht document)*

This chapter presents the main functions and operating parameters of a boring machine and states the different aspects that will be developed in the following chapters.

2.2. Different functions of a boring machine

All types of boring machines have the following functions in common:

– mechanized ground excavation and stabilization of the face,

– disposal of rubble (or mucking),

– monitoring and correction of trajectory,

– installation of pipelines by jacking.

They can be differentiated according to their method of mucking, done by mud circulation or using an endless screw creeper or by pneumatic suction.

2.2.1. *Mechanized excavation of the soil*

2.2.1.1. *Blasting the soil*

The head of the machine is equipped with a cutting wheel whose tools are used to blast the soil under the combined action of rotation and thrust. A crushing cone, located behind the cutting wheel and intended to reduce the size of larger elements to allow their mucking, is present on most machines. There exist different cutting heads for various types of soil (see Figure 2.2). They can be distinguished by their cutting tools and from the geometry of the wheel, particularly the size of the openings for clearing of the soil.

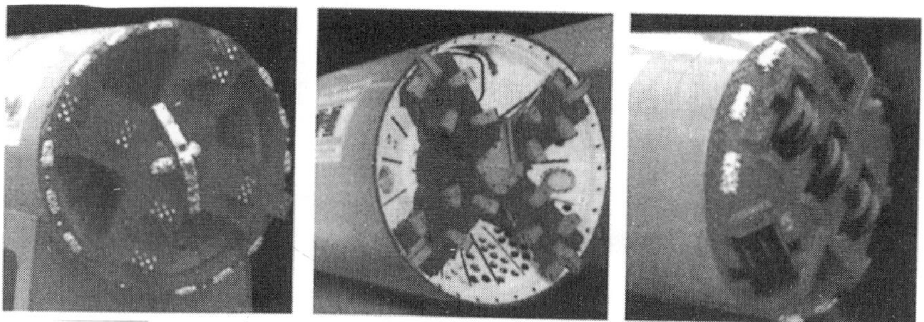

a) For sandy-gravely soil　b) For coherent soil　c) For rocks

Figure 2.2. *The different cutting heads (Herrenknecht document)*

For sandy or gravely soil (particularly alluvial), the cutting wheels are equipped with teeth (see Figure 2.2a). In rugged soil, these teeth dislodge the blocks, which are then crushed.

For coherent soil (silt, clay, marl), the cutting wheels are fitted with tools ("scrappers" or picks), which cut out chips of soil (see Figure 2.2b). On some machines, high-pressure water jets are sprayed on the wheels and in the stope to prevent sticking of clay and clogging of the mucking system.

Finally, for rocks (see Figure 2.2c), the cutting heads are equipped with rotary cutters having small openings. With the help of the thrust, the rotary cutters crush the rocks by means of shear and tensile stresses, which create cracks and loosen the fragments. These machines can bore through rocky soil with a compression strength of 200 MPa. This type of cutting wheel, also used in soil containing large blocks, is not suitable for clayey soil. Indications on the choice of the machines and equipment are given in Chapter 5.

2.2.1.2. *Confinement of the face*

To ensure the stability of the face, the contact pressure of the cutting wheel and the confining pressure must be equal to the earth pressure and to the pore pressure of water if the boring is done under the groundwater table. The total pressure thus applied on the head must be (see paragraph 3.1.4):

– greater than the active pressure of the earth so as to avoid over-excavation leading to the settling on the surface or even subsidence,

– less than the passive earth pressure so as to avoid forcing back the soil at the face, leading to elevations of the surface or lateral movements likely to create disorder for already-existing networks (Stein *et al.*, 1989),

In the case of hydraulic mucking, this pressure is ensured by the slurry injected into the chamber located at the back of the felling cone. It can be controlled more easily than the pressure exerted by the soil mixed in the stope of the screw type boring machines (Bennett *et al.*, 1994).

2.2.2. *Discharge of excavated earth (or mucking)*

2.2.2.1. *Hydraulic mucking*

Hydraulic mucking consists of removing the earth in suspension in a freely flowing fluid to the outside. In boring machines, water or pressurized bentonite slurry is injected into the stope (see Figure 2.3). This slurry is then pumped through a grill to a settling tank where the earth is separated from the mucking liquid. Mucking is monitored by regulating the rate of injection and discharge. These rates of flow must be suited to the nature of the soil, sufficiently high so as to avoid sedimentation of the earth and sufficiently low so as to prevent excessive erosion of the face.

If water can be frequently used as a mucking liquid, it is imperative to use slurry in soil that is not very coherent or very loose in order to ensure an efficient confinement of the face. In clayey soil, the discharge of earth by hydraulic mucking can be difficult due to sticking of material leading to clogging of pipes and the stope. In this case, it is convenient to use additives and/or regularly change the mud (or water) – see section 6.3.

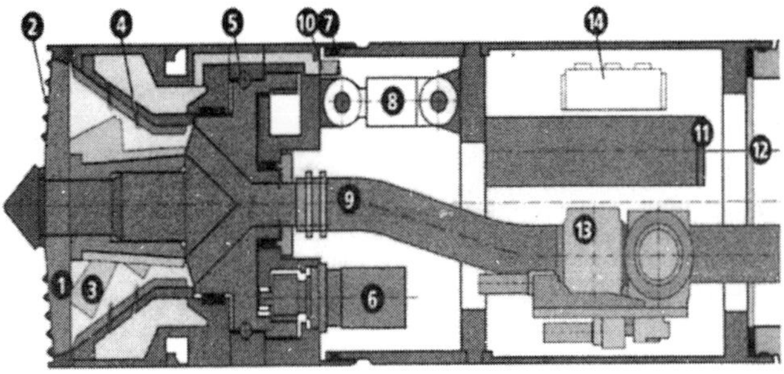

1. Cutting wheel
2. Tools
3. Crusher room
4. Injection nozzle
5. Main bearing
6. Cutting wheel drive
7. Coupling shaft
8. Thrust jack
9. Mucking pipe
10. Supply pipe
11. Target
12. Laser beam
13. Bypass
14. Valve block

Figure 2.3. *Principle of a hydraulic mucking boring machine (Herrenknecht document)*

Finally, special attention must be paid to the separation of rubble (see Chapter 6). In sandy or gravely soil, a simple settling pond is sufficient, but when the soil is clayey, the fines settle very slowly, which leads to a progressive saturation of the mucking liquid. A desander (hydrocyclone, vibrating bed) ensures the mechanical separation of solid particles from the mucking liquid, enabling it to be recycled for a longer period.

2.2.2.2. *Mucking with a screw conveyor*

The rubble is extracted from the stope using a spiral conveyor (see Figure 2.4). On some machines, the regulation of the rate of discharge of rubble, which controls the pressure of the earth at the face, is obtained by changing the rotation speed of the screw conveyor. The presence of a significant amount of sand, which risks getting set in the screw conveyor, or sticky clay, can lead to significant difficulties in mucking, thus making it necessary to add an additive in the form of a liquid or foam. Similarly, the elimination of big blocks can be a problem: a screw conveyor without a central hub enables the removal of blocks with a maximum diameter that is equal to 2/3 of the screw conveyor (Quebaud, 1996).

With a single motor to drive the wheel and the screw conveyor, the lengths of sections are limited to approximately 80 meters. A head driven independently of the screw conveyor can make longer sections while controlling the ground pressures in the stope.

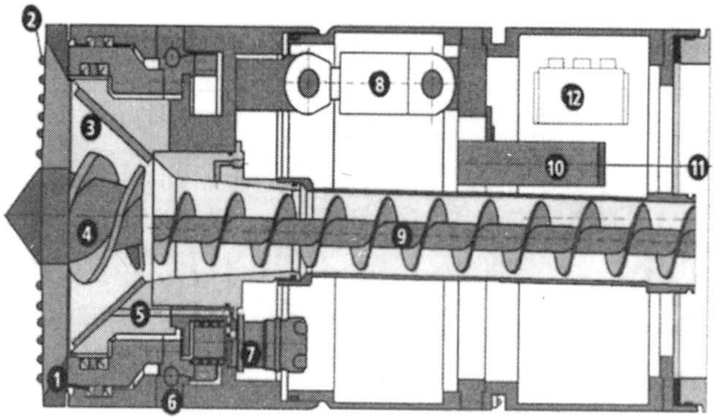

Figure 2.4. *Principle of a boring machine with screw conveyor mucking (Herrenknecht document)*

2.2.2.3. *Pneumatic mucking*

This is a system that is rarely used and consists of mucking by suction where the rubble is extracted from the face into an airtight vacuum container. The suction of the coherent soil is possible thanks to high-pressure water jets or compressed air injected directly into the cutting tools.

2.2.3. *Guidance and trajectory correction*

Controlling the actual trajectory of the boring machine, in relation to its theoretical position, is done using a laser beam with the sensor located in the start shaft whose impact on a target placed in the machine helps visualize the deviations in trajectory with the help of a camera onboard the boring machine. In this way, the horizontal and vertical deviations are monitored.

When the deviations become excessive, it is possible to correct the direction of the machine whose head is articulated by moving the cylinders located in the machine. There are usually three cylinders placed 120° apart, which allow the direction of the head of the boring machine to be corrected, both horizontally as well as vertically. This is a delicate operation, which will be discussed in greater detail in paragraph 3.1.2 and section 7.1, as it can have significant consequences on the success of a project.

This monitoring is generally done manually, but there are machines now that are equipped with an automatic guidance system using sensitive targets.

2.2.4. *Installation of pipelines by jacking*

Generally, the installation of a pipeline is done by successive jacking of pipes behind the boring machine (see Figure 2.1). This pipe jacking is ensured by a thrust frame equipped with hydraulic cylinders and located in the starting shaft. The machine advances with the help of the thrust transmitted by the pipes. The pipe elements generally have a length of one or two meters, or even three meters, with the dimensions of the shafts depending on this dimension. The capacity of the thrust cylinders varies from manufacturer to manufacturer and depends on the diameter of the boring machine: the thrust must be suited to the compression strength of the pipes and is controlled by the operator who can vary the travel speed of the thrust cylinders (see section 6.2).

An alternative method exists in inserting temporary steel pipes up to the exit shaft before jacking the final pipeline and then gradually removing the steel pipes. Another way is to replace the boring machine, after the first drive, by a reamer of larger diameter, which will be driven in the opposite direction from the exit shaft. After this, the permanent pipes are inserted; the temporary pipes are dismantled as they reach the starting shaft.

During direct jacking from the starting shaft, the length of the sections is limited by the resistance of the pipes and by the capacity of the thrust frame. Intermediary jacking stations are sometimes used to overcome these limitations. They are made up of hydraulic cylinders and a stress distribution ring fixed in a metal pipe inserted during sectioning. The alternating action of the thrust bench and the cylinders of the intermediary station make the pipeline "accordion". The total drive thrust is distributed over two or more pipelines if several intermediary stations are used. When the jacking is complete, the cylinders have to be dismantled in order to restore the internal continuity of the concrete coating of the sections. This dismantling has to be done manually and therefore the internal diameter must be sufficient (800 mm) for a localized intervention by an operator (Bennett *et al.*, 1995).

2.3. Various types of pipes

Pipes used in microtunneling must comply with the following standards:

– they must be made of suitable unit lengths,

– they must have a smooth outer surface and interlocking without flange to reduce friction,

– they must offer sufficient resistance to compression stresses during jacking,

– they must include an assembly system between pipes to enable transmission of the thrust stress while maintaining perfect water-tightness even in case of slight misalignments.

2.3.1. *Materials used*

Many types of materials are used: reinforced concrete (possibly with steel web), clay (vitrified or otherwise), glass fiber reinforced plastics (GRP) and steel:

– *concrete pipes* represent the majority of pipelines that are currently laid. For sanitary drainage, communication lines or electricity, the pipes are made of reinforced concrete manufactured by centrifugation. For pressurized installations such as water supply networks, special pipes made of concrete with steel web are used: these are pipes with a steel median ensuring water-tightness under pressure with double concrete coating. The resistance of concrete as well as the thickness of the pipes can be adapted for the thrust stress required for sinking. Use of "High Performance" concrete helps improve the resistance capacity to the thrust by about 80% in comparison with standard concrete,

– *clay pipes*, available in diameters of 150 mm to 1,200 mm, offer greater resistance than concrete pipes at the same thickness. When their surface is vitrified, it is extremely resistant to water absorption and chemical attacks. However, their manufacturing process provides high dimensional tolerance and makes it difficult to use lubricant injection nozzles often employed to reduce friction stresses,

– *pipes made of composite materials,* known as "glass fiber reinforced plastics" (GRP), offer very good resistance to corrosion and thus are particularly efficient in transporting corrosive fluids or for carrying chemically aggressive soil. Moreover, they offer high resistance at a lower weight. The external diameters available are between 400 and 2,400 mm. Their compressibility requires a clearance at the level of thrust cylinders during jacking over large distances (Boyce *et al.*, 1996),

– *steel pipes* have the major advantage of offering strong resistance (which provides the possibility of using smaller thickness), but they are sensitive to corrosion; in addition, joints between pipes are simplified: they can be directly bolted together.

Pipes made of asbestos cement were used when this technique was first being developed, but these have now been discontinued for health reasons.

2.3.2. *Joints between pipes*

The types of joints most widely used have the following common characteristics (see Figure 2.5):

– an outer ring of a diameter equal to or slightly less than that of the pipe, ensuring the self-centering of pipes and compression of the waterproofing material (1),

– a waterproofing device (seals in various forms) ensuring water-tightness from inside to outside and vice-versa (2),

– stress-spreading material limiting the concentration of stresses induced during jacking by applying off-centred stresses (3). Generally made of wood lath or dense bonded wood, the distribution of stresses at the ends of the pipes according to the angulation between joints depends on its thickness and mechanical behavior (*Pipe Jacking Association*, 1995). They are not used with GRP pipes.

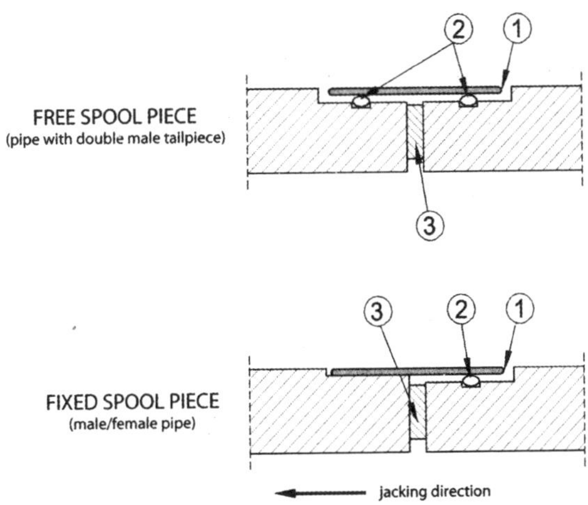

Figure 2.5. *Different types of joints*

2.3.3. *Resistance capacity of pipes*

It is mostly the resistance to thrust that determines the dimensions of microtunneling pipes. This resistance, of course, depends on the material constituting the pipes, but it also depends on the implementation characteristics: contact area and distributor material between pipes, right angulation of joints and off-centering of thrust cylinders.

According to the manufacturers, and depending on the material and thickness, the applicable thrusts for an internal diameter of 800 mm range from 850 to 3,000 kN. These Figures, however, need to be adjusted according to the driving methods accepted. As an example, and for concrete pipes, Figure 2.6 shows the variation of the acceptable total compressive force according to the angulation between two concrete pipes of 780 mm external diameter and for two joint thickness between pipes (3): we notice that it can drop by more than 50% when the angulation exceeds 0.6°; section 6.4 will present the methods enabling the evaluation of these effects in greater detail.

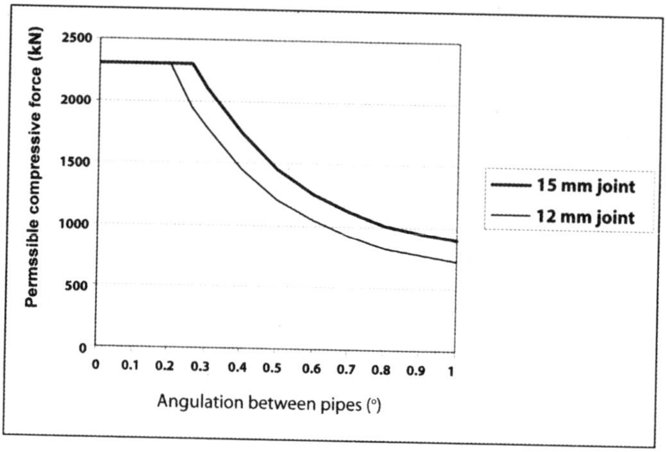

Figure 2.6. *Variation of acceptable force with angulation*

Chapter 3

Summary of Parameters Affecting Work at the Site

3.1. Summary of parameters affecting the microtunneling

The French National Research Project "Microtunnels" included in particular the continuous monitoring of fourteen microtunneling sections, of lengths varying from 40 to 170 meters and sunk at a depth of 1 to 30 meters. The nature of soils covered is of various types: ranging from clean fine sand to extremely soft clay including sand-gravel mixtures. Three types of boring machines were used, implementing different processes (hydraulic or pneumatic mucking, jacking in one or two stages).

The pipes installed at these sites all have a length of 2 meters and an internal diameter ranging from 500 to 1,000 mm. They are mostly made of concrete (BHP), sheet metal core concrete if the pipelines are for drinking water, and two of them are made of vitrified sand stones. The characteristics of each of these sites are given in Table 3.1.

Research done as part of the French National Research Project "Microtunnels" thus enabled us to characterize the ground crossed, to constantly record the digging and sinking parameters, and finally to also take note of the operator's choices (speed, guidance, etc.) and the construction parameters that are not recorded by the computer at the operator's console. They are at the heart of the current summary on the penetration parameters.

	Montmorency 2	**Châtenay Malabry**	**Neuilly**		**Bordeaux**			**Limoges**	
			1	**2**	**1**	**2**	**3**	**1**	**2**
Height of roof (m)	7	6	5	5	7	7	7	2 to 14	1 to 9
D_{ext} (m)	1.08	0.96	0.64	0.66	0.65			0.96	
Length (m)	170	166	113	95	90			98	95
Machine	Herrenknecht AVN 800 With reamer	Herrenknecht AVN 800	NLW		Markham 500			Herrenknecht AVN 800	
Mucking	Hydraulic	Hydraulic	Pneumatic		Hydraulic			Hydraulic	
Type of soil	Sand	Fine sand	Coarse sand and gravel		Fine sand			Weathered gneiss and rock sand	

a) Granular soil

	Montmorency 3	**Champigny**	**Barr 3**	**Barr III**	**Geneva**
cover(m)	4	5	4	3	7 to 30
D_{ext} (m)	1.08	0.75	0.97	1.178	1.275
Length (m)	120	80	60	40	170
Machine	Herrenknecht AVN 800 with reamer	Herrenknecht AVN 500 with reamer	Herrenknecht AVN 800	Herrenknecht AVN 1,000	Herrenknecht AVN 1,000
Mucking	Hydraulic	Hydraulic	Hydraulic	Hydraulic	Hydraulic
Type of soil	Loamy marl with gritter	Marl – clayey gravel	Sandy clay silt	Clay	Marl – sandstone

b) Cohesive soil

Table 3.1. *Sites monitoring during the PN*

3.1.1. *Rate of penetration*

The rates of penetration and their variation are essential data for the conception of the project. This paragraph therefore provides the average penetration values according to the nature of the soil and attempt to analyze the various parameters that influence the jacking rates, particularly in clayey soil.

We can characterize the rates of penetration of a boring machine by two different criteria: the jacking duration for a single pipe, which is determined by the nature of the soil and the digging technique, and the duration for the installation of a pipe, which additionally includes the time required for installing the pipes on the thrust ring and the making of various connections (mucking, electrical supply, etc.).

3.1.1.1. *Duration for pipe jacking only*

The penetration values obtained on the sections continually monitored during the French National Research Project "Microtunnels" were supplemented by the values recorded on nine other sections thanks to questionnaires filled in by several companies. The study is therefore based on a total pipeline of 1838 m of which 507 m are for cohesive soil, 855 m for fine sand and 476 m for sand and gravel. The analysis of all these results highlights the systematically different behavior depending on the three types of soil: marl/clay, fine sand, sand and gravel (see Figure 3.1).

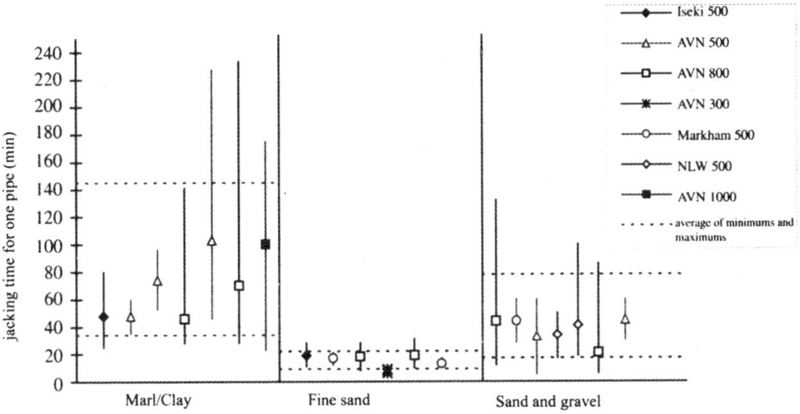

Figure 3.1. *Means, minimums and maximums of the durations of pipe jacking according to the nature of the soil*

We observe that the pipe jacking time in clay and sand and gravel is much greater than in fine sand. The penetration is directly linked to the crushing capacities of large elements and muck disposal capacities. In fact, in order to avoid creating over-thrust at the head, the penetration speed must be related to the rate of disposal of the excavation material. Similarly, problems relating to clogging of the tool and the grinder by clay, as well as the time required for crushing large elements, limit the jacking speeds.

3.1.1.1.1. Fine granular soil

In fine sand, the jacking rates are extremely uniform. The jacking duration for a 2 m tube varies very slightly around an average of 15 minutes (see Table 3.2 and Figure 3.1).

3.1.1.1.2. Coarse granular soil

In materials that are coarser, the speed of penetration is clearly lower. We have calculated an average jacking time of about 40 minutes. This value is quite variable as, depending on the sites, the jacking of a pipe takes on average 20 to 45 minutes. The lower jacking rate in sandy gravel soil is explained by the necessity of not exceeding the acceptable stresses for various systems at the head. The torque at the head increases according to the resistance of the ground, as well as the filling of the crusher by elements with high particle size requiring them to be crushed before being taken back by the mucking system. So that the crushing capacity is not exceeded and the torque at the head remains below the acceptable value fixed by the manufacturer of the machine, the operator of the boring machine is required to reduce the jacking speed.

	Average jacking time for a pipe (2 m)	**Total duration of the cycle (installation/connection/jacking/ maintenance)**
Fine sand	16 min. (2)	60 min.
Sand and gravel	38 min. (3)	90 min.
Clay/marl	70 min. (1)	120 min.

(1) Calculated over a straight section of 507 m; (2) Calculated over a straight section of 855 m; (3) Calculated over a straight section of 476 m

Table 3.2. *Summary of penetration rates for each type of soil*

3.1.1.1.3. Cohesive soil

The penetration rates in clayey soil are overall lower than those in granular soil. The great heterogeneity in jacking durations characterizes jacking in this type of soil. In fact, our site inspections have shown that a 2 m pipe was sunk in 70 minutes on average, but this jacking duration varied between 45 and 100 minutes.

The jacking rates are linked to the extrusion and evacuation capacity of the soil. The clay mixed with water becomes sticky and the various systems of the head (cutting wheel, mucking inlet sieve) may get clogged. In order to avoid creating over-thrust at the head and prevent the blocking of the cutting wheel, the jacking speed must be adapted to the rate of evacuation of the rubble.

3.1.1.1.4. Influence of the clay content

The strong disparity in the values recorded is obviously linked to the clay content of the soil and the proper adequacy of the digging technique with this kind of soil (adapted cutting wheel, transit of the muck in the crushing cone: see Chapter 5). To illustrate the difficulties encountered in the presence of clay, the following examples can be cited:

– the two microtunnels at Barr were sunk with similar equipment; however, the jacking duration was, on average, 1h40 per pipe on the Barr III section instead of 1h10 for Barr 3. The boring machine at Barr III was nevertheless better equipped with high-pressure jets, but the higher clay content (Ip = 35, against Ip = 19 for Barr 3) probably explains the greater clogging problems (FSTT RS22);

– at Champigny 4 (FSTT RS9 and RS14), we notice that, in spite of the stiff clay (IP = 19), the rate of jacking was just as bad (1h43). The heterogeneity of the soil (clay + block, alternate layers of clay/sand) did not allow the use of equipment specifically suited for clay: on the one hand, the use of a rock head with a small opening and equipped with rollers, required by the presence of blocks of large dimensions, proved to be unsuitable for cutting clay; on the other hand, the inability to modify the inlet point of the mucking liquid during jacking on the boring machine used did not allow the entire mucking liquid to be injected into the crushing cone, when this was required to prevent the head from clogging.

3.1.1.1.5. Impact of the quality of mud

It was noticed at all the sites built over clayey soil (FSTT RS9, RS14 and RS22) that the effectiveness of blasting and extraction of earth markedly reduced when the viscosity and density of the mucking fluid became too great. Inversely, the jacking speeds systematically increased when the mucking fluid that had become very thick and viscous due to fines from the soil was replaced by clean water.

In clayey soil, mucking with "clean water" seems to favor the erosion of the face and facilitate the passage of the excavated soil between the crushing cone and the confinement chamber. Depending on the clay content of the soil, the effect of mud renewal is more or less durable. Measurements of the viscosity of the mucking fluid at Barr 3 showed that 20 meters after renewal of the mucking mud, the viscosity had regained its value prior to renewal (FSTT RS22).

3.1.1.2. *Total duration for the installation of a pipe in the ground*

The total duration for the installation of a pipe includes, on the one hand, the installation of this pipe in the thrust frame of the shafts, the connection of mucking pipes and the different hydraulic or electrical cables required for controlling the boring machines and the various maintenance operations (adjusting the laser, draining the settling pond, etc.), and on the other hand, the jacking itself. On an average, over a complete section, the time required for operations other than jacking is about 40 to 45 minutes.

Thus, to install two meters of pipeline (a pipe) in the ground we can estimate, all operations taken together, an average of 1h00 in fine sand, 1h30 in sand and gravel and 2h00 in clayey soil (see Table 3.2). This last value can, however, be higher if significant clogging problems occur during jacking.

3.1.2. *Alignment deviations*

Respecting the tunnel alignment is an important factor that determines the success of microtunneling. First of all, in the case of the construction of a gravity system, it is indispensable to respect a precisely defined angle. In addition, significant deviations in trajectory leading to the misalignment of pipes with respect to one another may be at the source of stronger jacking thrusts on the one hand, and disorder in pipes, caused by concentration of stresses and loads during bending, on the other. We notice in general at the sites that the trajectory deviations may reach several centimeters, and they are usually more pronounced in the horizontal plane than in the vertical one.

3.1.2.1. *Human factors*

The guiding process is tricky and requires some experience. The boring machine consists of an adjustable head that is connected to a "fixed" body. Three cylinders located at the level of this mobile joint enable the head of the boring machine to be directed (see Figure 3.2).

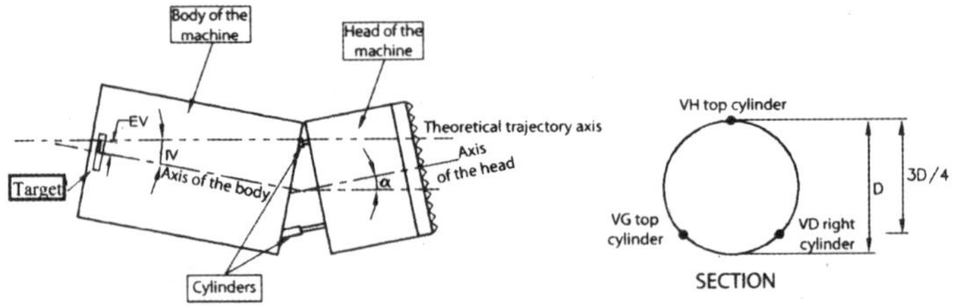

Figure 3.2. *Diagram of the steering cylinders of the boring machine head [FSTT RS22]*

When the machine veers off-course, the operator manipulates these cylinders in order to correct the angle of the head α and re-establish the trajectory. The operator has means to lengthen the steering cylinders and can thus apply the corrections in a controlled way.

However, there is a time lapse between the correction on the cylinders and the actual re-establishment of the trajectory, during which the boring machine continues to advance with a deviation in relation to its theoretical trajectory. The difficulty in controlling comes from the need to anticipate the action of the cylinders with respect to the current deviation, failing which the machine meanders following a sinusoidal curve and is difficult to control.

Figure 3.3, relating to the Neuilly site, presents the variation during jacking:

– of the vertical deviation with respect to the theoretical alignment, marked as EV,

– of the angle of the boring machine with respect to the theoretical alignment, marked as IV,

– of the angulation between the head α and the body of the machine (Figure 3.2).

This can be analyzed as follows:

– to correct the observed deviation, the operator must impose an angulation of the machine aimed at modifying its IV angle, a correction that was imposed towards pm 31,

– at pm 33, noticing that the trajectory of the machine is returning to its theoretical position, the operator significantly reduces the angulation. However, he reverses it only at pm 35, when the EV deviation is almost cancelled,

– this correction appears to be too late: we notice that the machine remains at a constant upward angle up to pm 39. This excessive-correction creates a new EV deviation in the opposite direction – one that is greater (45 mm) than the previous one. Thus, in this example, 6 meters were traversed between the correction by the operator on the steering cylinders and the change in the angle of the boring machine. While traversing these 6 meters, the boring machine continued to travel upwards, creating a new trajectory deviation. These delays in response, which vary depending on the nature of the soil, require great vigilance and experience on the part of the operator,

– only a strong angulation imposed after pm 39 helped correct the angle and return the machine to the theoretical trajectory.

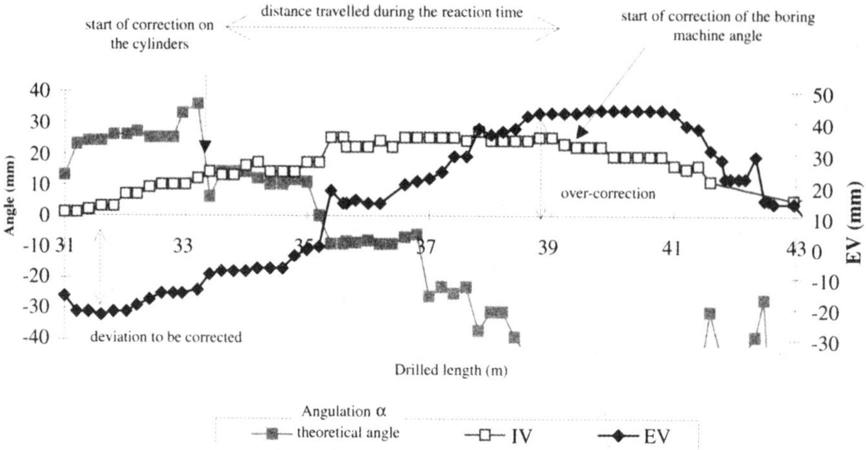

Figure 3.3. *"Over-correction" of the trajectory due to insufficient anticipation of the operator with respect to the response time of the boring machine (Neuilly 2, pm 31 to 43)*

Thus, in order to limit the amplitude of deviations, it is mandatory that the operator anticipates his action on the steering cylinders, taking into account the machine's response time that depends on several factors, particularly on the nature of the soil.

3.1.2.2. *Technological factors*

Figure 3.4 groups together the data obtained from the six sections monitored continuously. We have counted, by sections of drilled length, the proportion of vertical deviation peaks located in the following ranges: 5–10 mm, 10–20 mm, 20–30 mm, 30–40 mm and 40–50 mm. The sections not being of identical length, the number of sections concerned by the study is indicated in the form of a histogram.

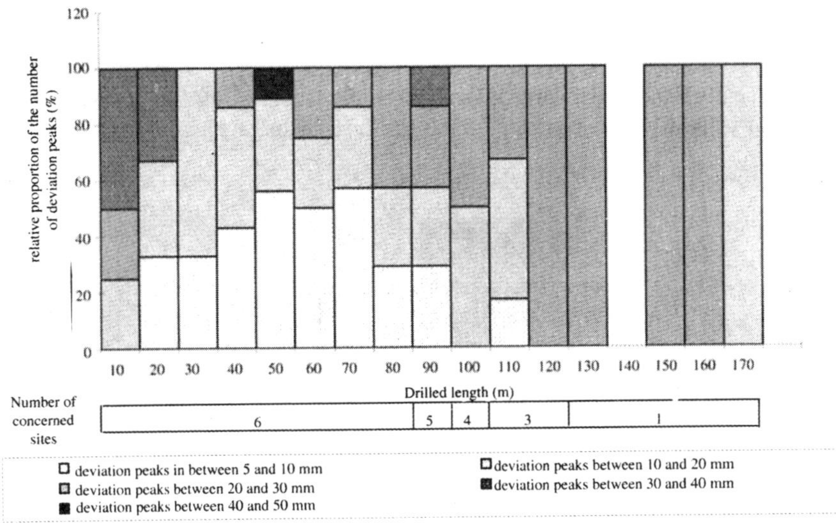

Figure 3.4. *Proportion of the number of vertical deviation peaks, per section of drilled length, at six sites*

We note that the largest deviations are at the start and end of the section: 50% of deviations are greater than 30 mm in the first 10 meters; deviations of more than 30 mm amplitude are mostly located in the first 20 meters. The analysis of sections monitored showed that it is an improper positioning of the thrust elements (downstream shell, thrust station, etc.) that causes these deviations of large amplitude at the beginning of sections (Pellet, 1997).

Between 20 and 70 meters, the deviation peaks greater than 20 mm represent only 15 to 25% of the total deviation peaks. Beyond this distance, the average amplitude of deviations clearly increases. The analysis of the six sections has highlighted two facts that may explain the greatest difficulties in guidance observed at the end of sections:

– from a certain drilled length the divergence of the laser beam makes the impact less accurate on the target,

– significant thrust strains recorded after a certain drilled length lead to deformations of the starting shaft, and as a result of the laser generator support,

– it is recognized in the documentation that a minimum overcut is required to be able to guide the boring machine. This is illustrated by the examples of Neuilly 1 and 2: both sections were cut in similar soil but the trajectory deviations recorded at Neuilly 2 are much greater; the lower width of the overcut on this section thus seems to be the cause of difficulties in guidance.

3.1.2.3. *Factors linked to the soil*

If we exclude the first 20 meters and limit ourselves to a maximum length of 80 meters, we note that the largest deviations are recorded on the two sections at Neuilly that were driven in coarse granular soil (see Figure 3.5).

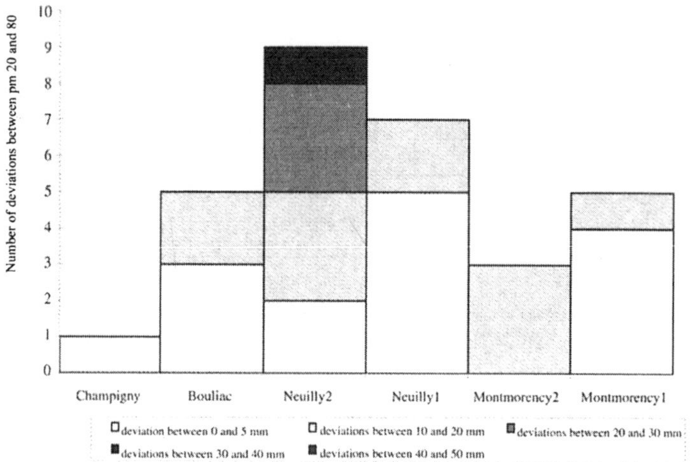

Figure 3.5. *Number of deviations per amplitude section, between pm 20 and 80 of six monitored sections*

The presence of coarse elements makes guidance more difficult as is shown by section no. 1 at Neuilly, for example (see Figure 3.6) [FSTT RS11]. Up to pm 70, in ground made up of sand, the trajectory deviations are low (< 20 mm). The amplitude of deviations clearly increases when the granulometry of the soil becomes coarser (sandy gravel + stones).

The analysis of response times of the boring machine in relation to the corrections of the operator shows that the fixed portion of the machine follows the change in angle of the movable head downwards less rapidly after pm 70. During the distance traversed, before the downward correction of the trajectory becomes effective, the boring machine continues to deviate upwards. This explains the greater upward deviations recorded after pm 70 when changing from sand to gravely sand, and then to gravel.

We can suppose that, in coarse soil, the peripheral tools ensuring the overcut favor the unearthing of large elements, which fall due to gravity into the crushing cone. The space thus created in the top portion of the face favors the deviations of the boring machine upwards.

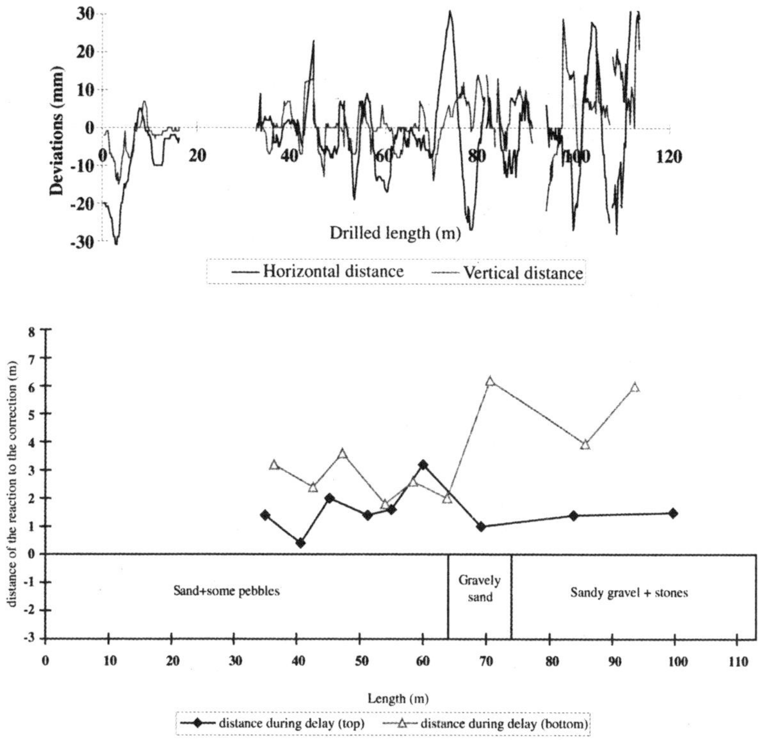

Figure 3.6. *Analysis of trajectory deviations of section no. 1 at Neuilly*

It is generally the sections in cohesive soil that present lesser alignment deviations. It is plausible that the improvement in the performance of excavation in cohesive soil makes the guidance process more effective.

3.1.3. *Frictional forces*

Frictional forces generally constitute the most important part of the drilling thrust. Increasing with the drilled length, it is these forces that actually limit the length of sections. It is therefore important to be able to accurately quantify them and analyze the parameters that influence their amplitude.

Experimental follow-ups undertaken during the National Project have helped illustrate the predominant impact of the overcut, the lubrication and the downtimes.

3.1.3.1. *Principle of analysis for experimental data*

The total thrust P consists of the thrust at the head Rp and the frictional forces F (see Figure 3.7).

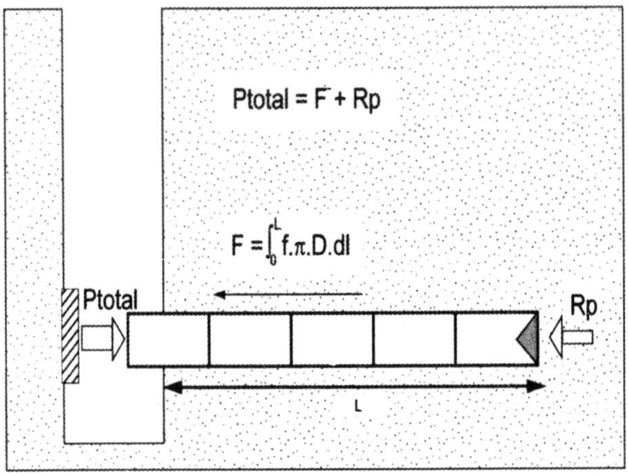

Figure 3.7. *Schematic diagram of jacking stresses*

The precise estimate of frictional forces assumes that we also know the total jacking thrust and the stress at the head. However, the stress at the head is not measured on most boring machines. During the two experimental follow-ups, the thrust at the head could be estimated thanks to special instrumentation (FSTT RS1 and RS11). In both cases, the measurements enabled us to establish that (see Figure 3.8):

– the local peaks of the total jacking thrust are linked, for the most part, to the radial cutting forces of the boring machine in the soil,

– the minimums of the total thrust correspond to a very low or even zero thrust at the head and can therefore serve as the basis for the estimation of frictional forces.

Thus, in the absence of thrust measurement at the head, we will estimate the soil-pipe friction curve from the envelope of the minimums of the total thrust curve; its gradient related to the drilled surface helps determine a value of the unit friction, having the dimension of a pressure. The analysis of these curves shows, moreover, that these minimums are encountered only during jacking, for the starting stages present, most of the time, higher thrust values. The friction deduced from the minimums of the thrust curves is therefore a *dynamic friction* (f).

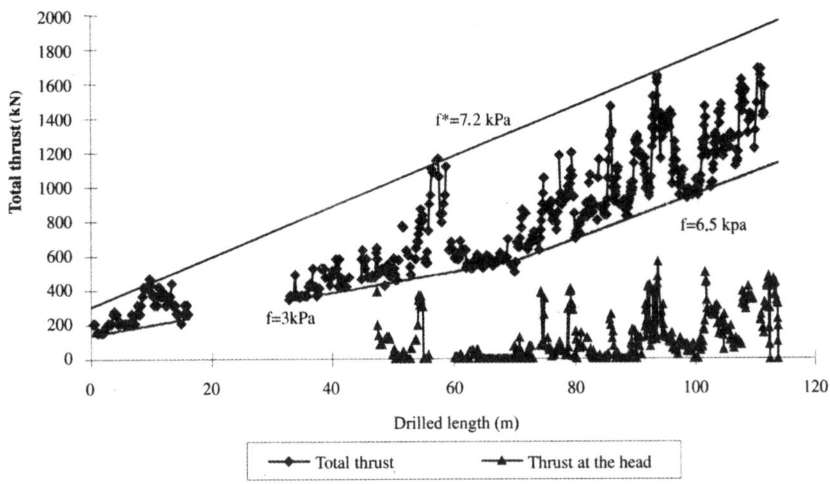

Figure 3.8. *Comparison of the variation of the total thrust and the thrust at the head at Neuilly 1. Evaluation of frictional forces*

The maximums of the total thrust that determine the possibilities of jacking over large lengths correspond either to stress peaks at the head or starting thrusts after an interruption in jacking: this phenomenon highlights the existence of *static friction.* We characterize these maximum stresses by the gradient of the envelope of their maximums which, related to the drilled surface, helps obtain an apparent coefficient of friction f* having the characteristics of a stress (see Figure 3.8).

Generally, after a certain drilled length, the thrust maximums correspond to the starting thrusts rather than the stress peaks at the head, and in this case the coefficient f* characterizes *static friction.*

3.1.3.2. *Effect of the overcut*

The impact of the overcut, i.e. the annular space between the pipes and the soil, on the frictional forces was clearly highlighted by the microtunneling at Neuilly 2 (FSTT RS11).

Figure 3.9 shows that after pm 16, pipes of larger diameter were installed, reducing the overcut from 32 to 12 mm. At the same time, the unit friction almost tripled, going from 3 kPa to 8 kPa then 10 kPa. This confirms the importance of the overcut, which by enabling radial decompression of the soil reduces the normal stresses acting on the pipe.

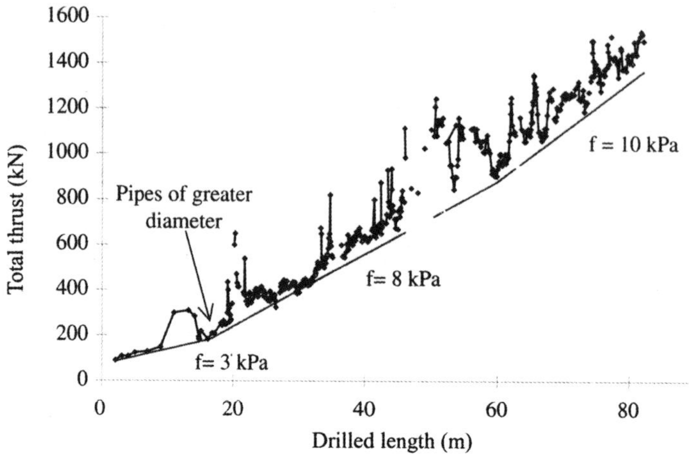

Figure 3.9. *Variation of the total thrust depending on the penetration at Neuilly 2. Impact of the overcut on the frictional forces*

3.1.3.3. *Impact of the downtimes*

Experimental follow-ups have highlighted, on certain microtunnels, starting thrust forces that are greater than those recorded prior to the stoppage in jacking.

This increase in thrust stresses during restart, i.e. frictional forces, if we assume that the thrust at the head remains the same, can be explained by soil creep, which leads to a tightening of the soil along the pipeline. It can also be due, in part, to the dissipation of induced interstitial overpressures in the bentonite film, which leads to the increase of the effective stress in the cake after draining of the bentonite, and as a result in an increase in frictional forces.

After a certain drilled length, the starting thrusts are systematically more penalizing. It is therefore necessary to be able to quantify the additional friction induced during starting (f_{sup}), which adds up to the dynamic friction applicable during jacking (f).

Increases in thrusts following interruptions in jacking (D/P = $P_{starting}$, – $P_{last\ thrust\ before\ shut\text{-}down}$) were recorded according to the penetration at the Champigny 4 site (see Figure 3.10). We have considered four main categories for stoppages: interruptions of less than 1 hour 30 mins corresponding to the setting up of the next pipe, stoppages of 1 hour 30 mins to 3 hours, from 14 hours to 20 hours (night) and stoppages of about 64 hours (weekend).

Increases in friction relating to stoppages of short duration (T < 3 hours) are difficult to analyze, as they are very short and mostly do not exceed the confidence interval for measurement. It seems, however, that the increase in friction caused by soil creep becomes sensitive, even for short downtimes, beyond 45 m of jacking.

For stoppages of more than 1 hour 30 mins, increases in the thrust at the start become much more significant, and we note a linear variation in these increases of thrust with the drilled length overall.

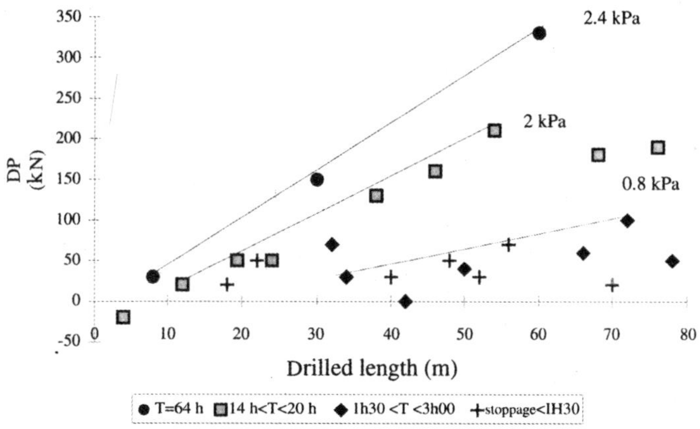

Figure 3.10. *Champigny 4, increase in the thrust following jacking interruptions, according to the drilled length*

Over the entire section, the increase in the unit friction caused by the tightening of the soil is of:

- 2.4 kPa for stoppages of two and half days (T = 64 hours),
- 2 kPa for everyday stoppages (14h < T < 20 hours),
- 0.8 kPa for stoppages of short duration (T < 3 hours).

Thus, the additional duration that adds up to the dynamic friction during jacking depends on the downtime. We have shown a linear relationship between the relative increase in the jacking thrust [$(DP/P_{before\ stoppage}) \times 100$ in %] and the logarithm of the downtime expressed in hours (Pellet, 1997). The gradient of the increase in thrust with the logarithm of the downtime varies between 6 and 8.

These values, however, need to be considered with caution because of the significant dispersion of points for stoppages of short duration. On the sections at Champigny, Montmorency 2 and 3 and Bouliac, the additional friction resulting from a stoppage of less than 3 hours is between 0.6 and 0.8 kPa, and between 1.1

and 2 kPa for stoppages of about one night (see Table 3.3). The sections at Neuilly, drilled in coarse soil and with low injection of bentonite, present a very slight increase in the starting thrust.

	Champigny 4	**Montmorency 2**	**Montmorency 3**	**Bordeaux**
Nature of the soil	marl and gravelly-sandy marl	fine sand not very clayey	silty marl at times a little sandy	clean sand
Average rate of lubrication	26 l/ml	107 l/ml	95 l/ml	168 l/ml
Overcut (tunnel borer/pipe)	10 mm	30 mm	30 mm	10 mm
(D/P/P)%=p*ln(T)+q	8.2*ln(T) + 2	5.7*ln(T) + 11 (pm 60 to 122)	7.6*ln(T) + 8 4.8*ln(T) + 3 (**)	no stoppage > 3h
additional friction 12 h <stoppage< 30 h	2 kPa	1.1 kPa (pm 60 to 122)	1.3 kPa 0.6 kPa (**)	no stoppage > 3h
additional friction 1h30<stoppage< 3h	0.8 kPa	(pm 60 to 122) 0.7 kPa	0.6 kPa	0.7 kPa (pm 66 to 90)
f	2.8 kPa	2 kPa (pm 60 to 122)	3.3 kPa	1.7 kPa
$(f.\ add_{12<T<30}/f)*100$	70%	55%	20% (**) to 40%	

(**) Relationship established on sections where the lubrication was significant

Table 3.3. *Impact of jacking stoppages on the increase of frictional forces during starting*

These results show that prolonged stoppages may be the cause of a significant increase in friction. In the most unfavorable cases, the increase in thrust upon resumption of jacking can reach 70%.

It was thus observed that the amplitude of this additional friction could be reduced by sufficient lubrication. This was particularly appreciable on sections at Montmorency 2 and 3. At Montmorency 2, we find a reduction in the additional friction over the two sections (pm 20 to 45 and beyond pm 122) where we observe a greater volume of lubricant injected per linear meter drilled (see Figure 3.11).

The impact of lubrication on the amplitude of additional friction will be expanded in greater detail in the following paragraph.

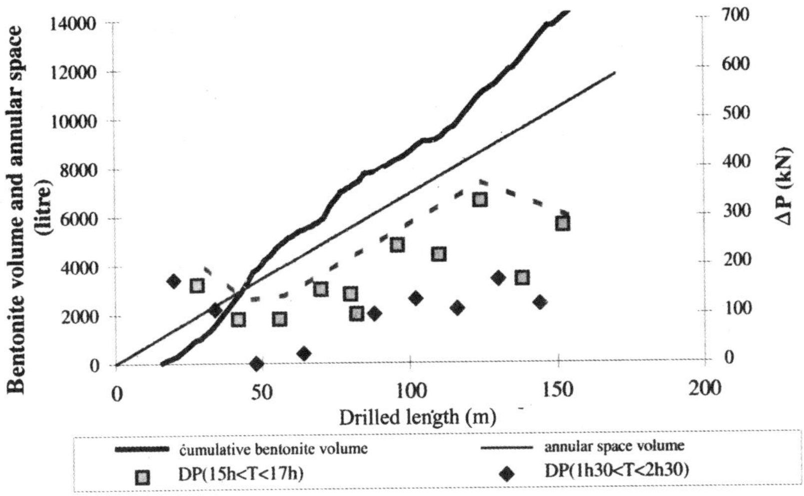

Figure 3.11. *Impact of the volume of bentonite injected in the annular space on the increase in starting thrust*

3.1.3.4. *Impact of lubrication*

The difference in diameter between the boring machine and the pipes sunk leaves an annular space (known as overcut) into which a lubricating grout is injected in order to limit the frictional forces between the soil and the pipes.

The lubricant used at all monitored sites consist of a bentonite grout, combined sometimes with polymers and microbeads, such as at Bouliac for example, which increases the viscosity of the mud and improves its lubricating power.

At the monitored sites, except for the Bouliac section, the lubricating mud is injected only at the back of the trailing pipe through three points located at 120° on the circumference.

At Bouliac, during the first jacking stage, it is not concrete pipes that were sunk, but temporary steel pipes of which three were equipped with injection valves. Thus, the lubricating grout could be injected alternately in three different spots during jacking: at the level of the trailing tube and the tubes no. 8 and 20, i.e. at 4 m, 20 m and 44 m of the face. The mud was injected at the top of the pipes.

Table 3.4 presents the various sections that were subjected to lubrication. We will examine below the different observed effects of this lubrication.

3.1.3.4.1. Reduction in dynamic friction

On all the monitored sites, it was seen that the lubricant helped considerably reduce dynamic frictional forces. As for technical reasons the lubricant is injected only after 20 to 30 meters of jacking, we were able to easily quantify the impact of the latter on frictional forces.

For example, for microtunneling at Chatenay-Malabry (Figure 3.12), lubrication led to a reduction in unit friction from 7.3 kPa to 1.7 kPa, i.e. a reduction of 77% with respect to the friction without lubrication.

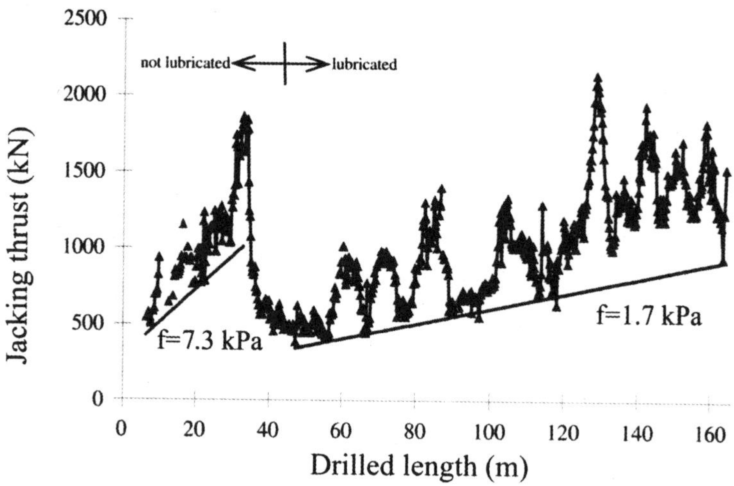

Figure 3.12. *Variation in the jacking thrust at Châtenay-Malabry*

	Montmorency 3	**Champigny 4**	**Geneva**	**Châtenay -Malabry**	**Montmorency 2**	**Bouliac 3**
Nature of the soil	Silty marl, a little sand in places	Marl and gravely-sand marl	Sandstone marl and altered sandstone	Fine clean sand	Fine sand not very clayey	Clean sand
Length of section (m)	120	80	170	166	170	90
External diameter of pipes (mm)	1,080	760	1,275	960	1,080	650
Overcut (mm)	30	10	20	15	30	10
Nature of lubricant	Bentonite	Bentonite	Bentonite	Bentonite	Bentonite	Bentonite + polymer
Injection points	4 m from the face 	4 m from the face 	4 m from the face 	4 m from the face 	4 m from the face 	4 m, 20 or 44 m from the face 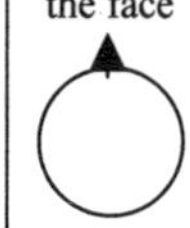
Volume of lubricant injected per meter sunk	95 l/ml	26 l/ml	40 l/ml	?	107 l/ml	168 l/ml
V. lubricant/ V overcut	1.38	0.72	1	?	1.55	5.6
Start of injection	pm 32	pm 22	pm 2	pm 34	pm 16	pm 16

Table 3.4. *Presentation of various sections monitored and the conditions for the injection of the lubricant*

When we consider all the monitored sites, we note that the reduction in frictional forces, following the injection of the lubricant in the annular space, varies between 45% and 90% (see Table 3.5).

	Montm. 3	Champigny	Châtenay	Geneva	Montm. 2	Bouliac
f (kPa) without lubrication	5.8	5.3	7.3		5.2	4.7
f_{lub} (kPa) without lubrication	3.3 average	2.8	1.7	0.65-2.32	1.9	0.5
$(f - f_{lub})/f$ Overall reduction in friction	43% average	47%	77%		64%	89%
Av. Rate of injection: vol. bentonite (l/ml)	95 average Loc[t] 180-250	26		40	107	168
Average rate of injection: V bento/V. overcut	1.38 average Loc[t] 2	0.72		1	1.55	5.6
f_{sup} (Pa) additional friction 12 h < T < 30 h	1.3 average Loc[t] 0.6	2.0		1.0	1.1	No stoppage

Table 3.5. *Impact of the injection of lubricant on the dynamic friction and additional friction following a stoppage for about one night*

Milligan and Marshall (1998) carried out a similar study over six sections and found a reduction in friction of the same order of magnitude, ranging from 45 to 95%. These two studies confirm that the injection of bentonite slurry (with or without a polymer additive) has a significant impact on the amplitude of frictional forces between the soil and the pipes.

3.1.3.4.2. Reduction in additional friction after immobilization

A comparison between the different sections where we were able to estimate the additional friction following stoppage for one night, as well as the analysis of the variations in the quantity of grout injected at the Montmorency 3 section, shows a clear reduction in the amplitude of additional friction as the quantity of grout injected is increased. The additional friction, equal to 2 kPa for an injection ratio of 26 l/ml, drops to 0.6 kPa when this ratio increases to about 200 l/ml (see Table 3.5).

3.1.3.4.3. Summary on the effectiveness of the lubrication

The effectiveness of the lubrication seems to be related to the quantities injected, the injection method, the nature of the soil and finally the type of lubricant.

Thus, we have observed a strong correlation between the injected quantities and the reduction in dynamic friction, ranging from 45 to 90% when the injected volumes vary from 25 to 170 l/ml (see Table 3.5).

Figure 3.13 presents the relative reduction in friction in relation to the volume of grout injected per linear meter sunk, by considering the overall estimated value over the entire section as well as the values recorded over various sections of the linear portions that differed from one another in terms of the quantity of lubricant injected.

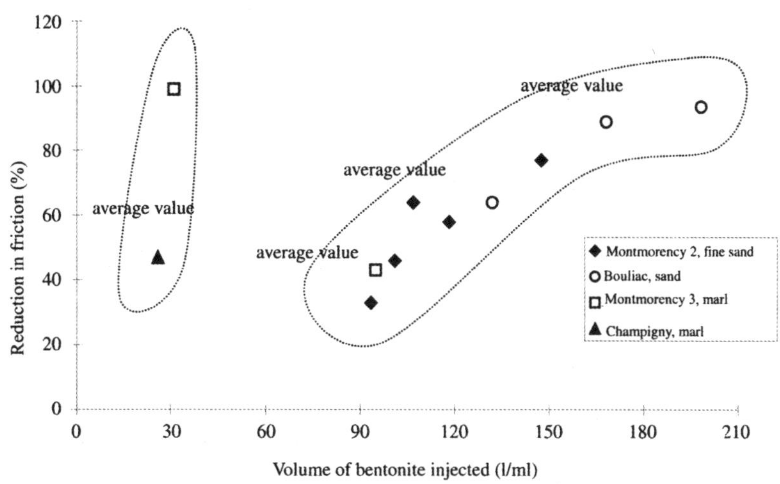

Figure 3.13. *Comparison between the percentage of reduction in the unit dynamic friction and the volume of bentonite injected per linear meter*

We observe a greater effectiveness in the injection of grout in marl (Champigny 4 and Montmorency 3 for the section injected continuously) than in sand (Montmorency 2 and Bouliac). It therefore seems that in soil without cohesion and permeable soil, a larger quantity of bentonite grout may be necessary to obtain the same reduction in frictional forces than that obtained in soil that is not so permeable.

In fact, the importance of the penetration of the bentonite suspension in the soil is, among other things, a function of the permeability of the soil. Following the injection of bentonite in the annular space, the soil around the pipes progressively seals off by the filling of pores, then by obstruction of these by solid particles in the suspension. Subsequently, when the suspension is thus blocked in the pores, there occurs a phenomenon of filtration and the solid part of the suspension forms a cake, comparable to a waterproof and resistant membrane. The greater the permeability of the soil, the more the suspension disperses in the soil and lower the quality of the cake formed.

Kanari (1998) similarly showed a relationship between the permeability of the soil and reduction in friction: the reduction in friction is 30 to 80% for a permeability coefficient less than 10^{-6} m/s; it is only 20 to 30% when the permeability varies between 10^{-4} m/s and 10^{-5} m/s, and it is zero when it reaches 10^{-3} m/s.

The values recorded in sandy soil show a close link between the volume of grout and reduction in friction. The quantity of grout seems to compensate for the largest seepage of the latter in the sand.

In addition, regarding the injections, we note:

– a drop in effectiveness when the injections are carried out discontinuously, as a reaction to the increase in frictional forces,

– the beneficial effect of the injection of lubricant done at various points of the pipeline, to maintain good lubrication over the entire section sunk.

Finally, the good results obtained at the Bouliac site following the attention paid in lubrication and reduction in friction need to be emphasized:

– addition of polymers and microbeads to the bentonite slurry,

– continuous injection at several points of the pipeline,

– importance of injected volumes (average injection rate of 5.6 times the volume of the overcut),

– jacking without long stoppages.

All these measurements helped maintain the friction at a particularly low level, of the order of 0.5 kPa.

3.1.3.4.4. Analysis of the soil-grout-pipes interaction

The models normally used to calculate the frictional forces between the soil and the pipe are based on three configurations:

– either the soil can close around the pipeline; the normal forces are then determined from patterns based on the silo theory proposed by Terzaghi,

– or the drilling remains stable and the convergence is less than the free space; the pipe then remains at the excavation invert and the frictional force is proportional to its weight,

– finally, in the last scenario, if the annular space is filled with bentonite slurry, it is advisable to take into account the weight of the pipeline without water. If this becomes negative, the pipeline will float and the friction will take place along the crown of the excavation.

We have observed that an injection of sufficient bentonite helped limit the additional friction following stoppages in jacking. We can thus assume that the grout injected between the ground and the pipes acts as a supporting and stabilizing agent for the excavation, preventing the tightening of the soil around the pipeline.

Furthermore, the frictional forces measured when bentonite grout is injected are, at the Bouliac, Geneva and Montmorency 2 sites, very close to the friction calculated considering only the weight of the pipes, i.e. considering that the excavation is stable and the pipeline slides on its base (see Table 3.6). Before injecting the grout in the annular space, the values of frictional forces recorded were close to those calculated using the Terzaghi model, i.e. by calculating the stresses that were applied on the pipes due to the weight of the soil and the arch effect.

We can thus deduce that the injection of bentonite grout at these three sites helped stabilize the excavation. It is possible that, on the one hand, the resistance to shearing of the bentonite grout provides additional cohesion to the soil and that, on the other hand, the formation of a watertight and resistant cake constitutes a veritable "shield" that enables the slurry pressure to exert a containment opposing the thrust of the soil.

On the contrary, at Champigny 4 and Montmorency 3, the friction values obtained are higher than the values calculated based only on the weight of the pipeline. At Champigny the quantity of grout injected was not enough to fill the entire annular space (see Table 3.5), and at Montmorency 3, the injection was done highly discontinuously. At both these sites, and unlike Bouliac, Geneva and Montmorency 2, the pipeline was still subjected to the action of the soil.

	Bouliac	**Geneva**	**Montmorency 2**
Soil	Clean sand	Marl and altered sandstone	Fine sand not very clayey
Friction measured without lubrication	4.5 kPa		5.2 kPa
Friction calculated using the Terzaghi model	5 to 5.4 kPa		8 to 8.8 kPa
Friction caused by the pipes' own weight	0.5 kPa	0.65 to 2.32 kPa	1.9 kPa
Friction caused by the pipes' own weight	0.5 kPa	0.8 kPa (0.6 kPa if the pipes are floating)	1.2 kPa

Table 3.6. *Comparison of actual friction values with the friction values calculated by supposing the stability of the excavation or otherwise*

3.1.3.5. *Impact of misalignment*

Various authors have observed an increase in the total thrust in relation to the deviations (Guilloux and Legaz, 1992; Milligan and Norris, 1999; Pellet, 1997).

Thus, there seems to be a clear link, on the two sections studied as part of the National Project (Montmorency 2 and Châtenay-Malabry), between the strong local increase in thrust (up to 500 kN at Châtenay and 250 kN at Montmorency 2) and the alignment deviations (FSTT RS1 and RS14). Similarly, the correlations between increases in thrust and horizontal deviations were observed at several sites (Guilloux and Legaz, 1992).

Measurements using strain measurement on concrete pipes at Châtenay-Malabry showed that sudden increases in thrust can be explained by greater frictional forces on the body of the machine, following trajectory corrections imposed by the operator (see paragraph 3.1.4.3).

Milligan and Norris (1999) also measured, thanks to the instrumentation of certain pipes, local increases in the contact pressure at the soil-pipe interface during trajectory corrections. Moreover, they were able to show, in the only case where the excavation remains stable and where consequently the pipeline is laid while being placed at the bottom, an increase in the frictional forces following horizontal deviations in trajectory, with the vertical deviations having no impact. The ascertained increase reaches 20 to 100% depending on the radius of curvature of the deviation and the length of the section.

3.1.3.6. *Impact of granulometry*

On section no. 1 at Neuilly, following the change from sand to gravely sand and gravel, the unit friction increased from 3 kPa to 6.5 kPa (Kastner *et al.*, 1996). The coarser the granulometry of the soil, the greater its dilatancy; the impact of the overcut is therefore less favorable, which leads to higher friction values. Studies undertaken by the JSTT (*Japan Society of Trenchless Technology*) (JSTT, 1994) also recorded systematically higher unit frictional forces in sand and gravel than in other soil.

3.1.4. *Stresses at the head*

3.1.4.1. *Presentation of general results*

During jacking, the head of the boring machine exerts a contact pressure on the soil. This force can constitute a non-negligible portion of the total thrust and it needs to be taken into account, like frictional forces, during the sizing of a microtunneling project. In addition, the amplitude of the stress at the head determines the stability of the soil at the face. If this is lower than the active pressure (thrust pressure), the soil can collapse, thereby risking the creation of subsidence. On the contrary, if it becomes greater than the passive pressure (stop), the head of the boring machine will pierce the soil and cause significant reverse movement in the block, as well as upheavals at the surface.

Generally, the thrust at the head R_p is not directly measured during jacking. It can however be evaluated by subtracting the dynamic frictional forces F estimated using the method presented in paragraph 3.1.3.1 from the total jacking thrust P_{tot}.

$$R_p = \frac{P_{tot} - F}{\pi.D_e{}^2/4}$$

This method does not distinguish between thrust peaks caused by static friction during restart and the increase in the thrust at the head.

If we consider all the sites monitored, we obtain an average thrust at the head which varies very little with the nature of the soil: it is between 180 and 350 kPa for clayey soil (average = 260 kPa) and between 200 and 540 kPa for sandy and sandy-gravel soil (average = 320 kPa) (see Figure 3.14). On the other hand, the maximum thrust is twice greater on average in sand and sandy-gravel (1,330 kPa) than in clay (600 kPa); in addition, the values of the thrust at the head in granular soil vary significantly (see Figure 3.15).

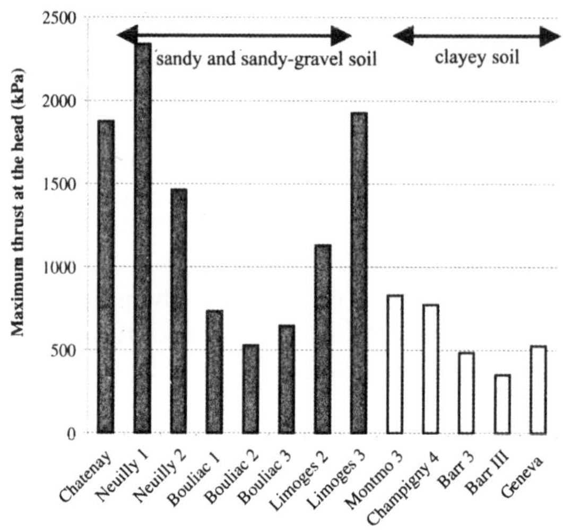

Figure 3.16. *Maximum thrust at the head*

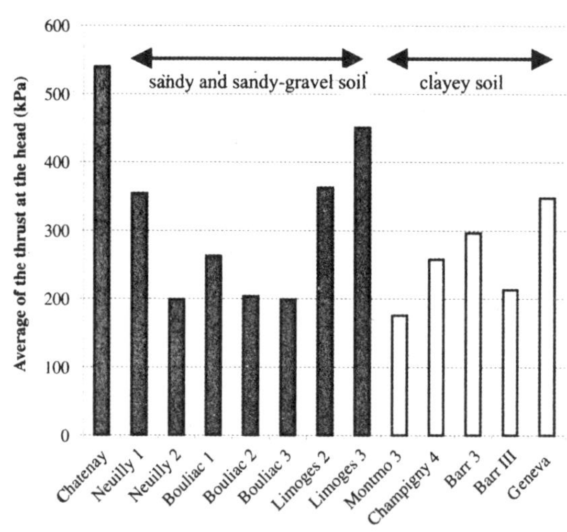

Figure 3.15. *Average thrust at the head*

We can also quote the guided study of the Working Group 3 and the JSTT (1994) on about 200 sites constructed by boring machines with hydraulic mucking. This estimates the thrust at the head from the value of the initial jacking thrust. They have shown a linear relationship between the external diameter of the boring machine and the thrust at the head, all types of soil taken together:

$$R_p \text{ [kPa]} = 650 - 0.3D_{ext} \text{ [mm]}$$

For external diameters covered by our study, the thrusts at the head will vary between 320 and 450 kPa, which correspond overall to the order of magnitude found here.

3.1.4.2. *Influence of blasting and mucking*

The most detailed analysis of the thrust at the head over several sites (FSTT RS 11 and RS 22) showed that the increases in thrust at the head corresponded almost systematically to increases in the torque at the head, i.e. to the power enabling the rotation of the cutting wheel at a certain speed (see Figure 3.16).

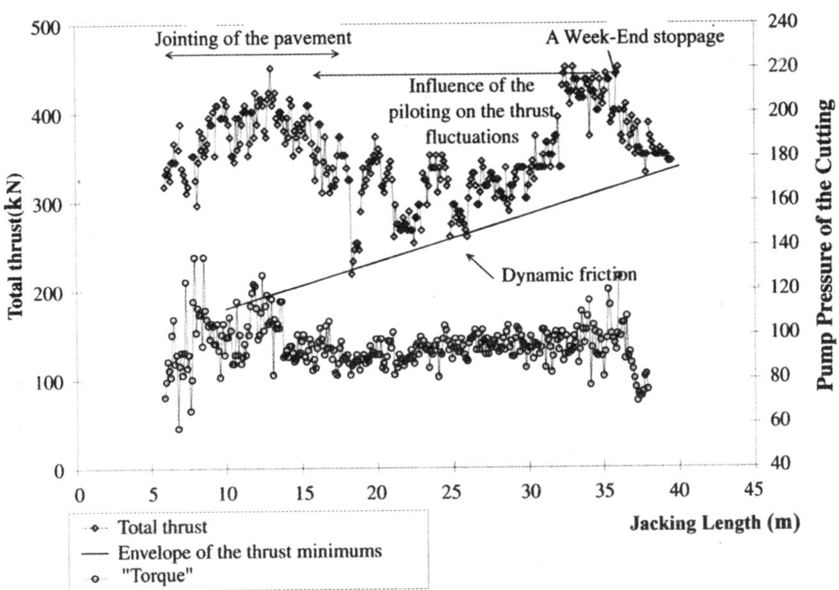

Figure 3.16. *Evaluation of the total thrust and the "torque", Barr III site [FSTT RS 22]*

The torque depends on the ease with which the wheel can cut and extract the material at the face and the capacity of grinding coarse elements and then swallowing them. When difficulties arise in grinding or clogging by clay at the level of the crushing chamber, and the jacking speed is not reduced as a result, there is a tamping at the head and thus an increase in thrust.

In sandy and sandy-gravel soils, increases in the thrust at the head seemed to be essentially linked to the difficulties in grinding during the crossing of extremely coarse ground, as was the case at Neuilly (old alluvions) and Limoges (altered gneiss) (FSTT RS 11 and RS 24). However, significant increases in thrust at the head were also noticed following reductions in the mucking flow. In highly permeable ground, such as at Bouliac, loss of mucking fluids in the ground reduced the effectiveness of the removal of earth and caused "tamping" of the crushing chamber, then an increase in the thrust.

In clay, the wheel and the grinder may clog in the case of a difficulty in removing the earth. In fact, clay mixed with water becomes sticky and the improper removal of the earth from the chamber as well as problems of mucking can cause tamping of the head. If the speed is not reduced as a result, the jacking is then done by partial forcing back, accompanied by a strong increase in the thrust at the head as noticed at Barr III (see Figure 3.16). At this site, the forcing back of the soil was visually translated at the surface by the fissuring of the pavement. The calculations based on the limiting equilibrium method seemed to indicate that the thrust values at the head had exceeded the stop value of the soil. From pm 15 onwards, a reduction in the jacking speed by the operator helped reduce the thrust at the head and did not create any disturbance at the surface.

3.1.4.3. *Influence of trajectory deviations*

Sometimes we observe ad hoc increases in the total thrust that do not correspond to an increase of the thrust at the head (as they are not accompanied by an increase in torque), but they coincide with changes in direction of the machine body.

Thus, ad hoc increases of more than 500 kN (670 kPa) at Châtenay-Malabry coincide with significant vertical deviations (see Figure 3.17). The recording of deformations by extensometric gauges on the second pipe after the head shows that the source of these sudden increases in the total thrust are essentially located at the level of the boring machine.

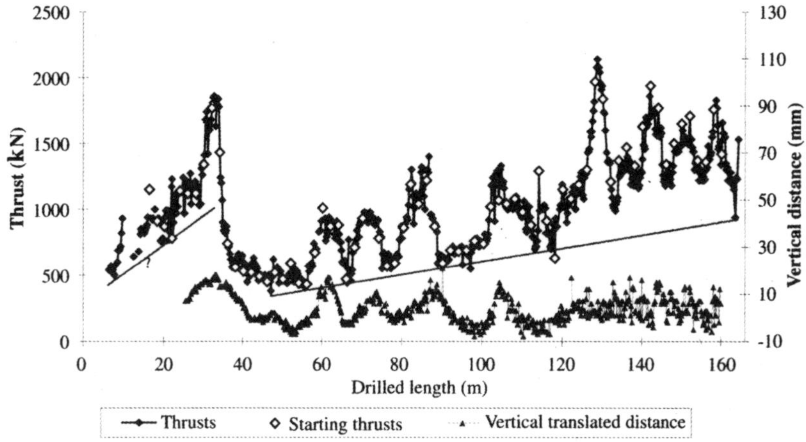

Figure 3.17. *Comparative change in the total thrust and the vertical deviation according to the drilled length, Châtenay-Malabry site*

It is possible that, when re-establishing the alignment followed by the body of the boring machine according to the correction imposed by the operator, the frictional forces on the 3 m-long non-articulated portion of the boring machine increase significantly. Once the fixed body of the boring machine has changed direction, the 2 m-long concrete pipes with a diameter that is less than that of the boring machine provide less resistance during passage of the bend, and the frictional forces reduce as well. Similarly, at Montmorency 2, ad hoc increases of about 250 kN (260 kPa) of the total thrust coincide with the changes in direction of the machine body.

3.2. Description of the main hitches that can occur when constructing a microtunneling site

3.2.1. *Blocking of the machine*

3.2.1.1. *Various boulders and obstacles*

The microtunnels projects are situated most often in the surface section of the ground, generally in between 3 and 15 meters deep, where we frequently come across loose soil. This loose soil (man-made backfills or mantle rocks) is likely to contain natural or surface heterogeneities.

Backfills in urban sites may contain demolition and rubbish material, old foundations, neglected piles or shafts, not to mention the various abandoned or functioning conduits and networks. Moreover, certain mantle rocks sometimes contain boulders of hard rock; this is particularly the case at the base of alluvia and within moraines.

The immobilization of the machine due to the encountering of obstacles may have various reasons:

– the boulders and obstacles whose diameter is too large to enter the crushing cone of the machine have to be broken by tools of the cutting wheel. This involves the use of a "rock" type wheel, with rollers and having very few openings in order to bear a significant stress at the head. The resistance to compression of the blocks that can be crushed with a "rock" head varies according to the power available at the head and according to the stresses permissible by the various systems of the head (drive shaft, steering cylinder, etc.). Most manufacturers claim that rocks with a compression strength of 100 MPa can be easily drilled through, and that the isolated blocks whose compression strength exceeds 200 MPa can also be gotten through. In addition, the compactness of the matrix in which the blocks are anchored must also be taken into account, as if these are not "held" properly, they could be pushed by the machine instead of being crushed;

– the various blocks or debris that enter the cone have to be reduced to a size that is suitable for their removal by the mucking system. And yet, the crushing capacities of the crusher are limited:

- by the size of the blocks: the diameter of blocks must be less than 30% of the inner diameter of the machine,
- by the quantity of material that can be crushed simultaneously,
- by the resistance of blocks: the blocks whose compression strength exceeds 100 to 200 MPa (for isolated blocks) cannot be crushed in the cone; however, it is difficult to state a limiting resistance, as this depends on the first two parameters,

– pieces of scrap metal, wood or PVC are difficult to crush by the crusher and the cutting wheel due to their flexibility. In addition to the risk of immobilisation of the cutting wheel by pieces of large size, small pieces can also cause difficulties in evacuation by blocking the inlet mucking grille or by forming clusters in the mucking system (see "sticking of clay", paragraph 3.2.1.4). The accumulation of muck in the cone can cause clogging which can lead to exceeding the capacity of the machine's crusher, or even a blocking of the machine requiring specific intervention.

3.2.1.2. *Excessive friction*

The drilling thrust is made up of the thrust at the head and the frictional forces along the pipes, which increase with the drilled length. The latter constitutes the predominant portion of the drive thrusts and is sometimes responsible for a premature break in jacking. In fact they can, after a certain length, exceed the thrust capacities of cylinders or the resistance of pipes and cause the microtunneling to be stopped (FSTT RS 11).

As we have shown in paragraph 3.1.3, the value of the frictional forces is highly variable; the soil-pipe friction is determined by the soil, the execution techniques (overcut, lubrication) as well as boring factors (duration of interruption in jacking and trajectory deviations). Because of this, there are no simple theoretical laws to estimate them. Section 6.2 will provide the guidelines for their evaluation.

Certain hydrogeological and geotechnical situations unfavorable from the point of view of frictional forces require increased vigilance:

– in permeable ground subjected to circulations of water (sand under the water table, highly cracked soil, etc.), the injected lubricant and the mucking fluid may disperse in the ground and not fill the annular space,

– in unstable, strongly convergent, bulky or dilating soil, the ground can tighten rapidly around the pipes,

– blocks destabilized by the excavation can furthermore create a buttressing stress on the pipes.

After a period of immobilization, the resumption of jacking requires a thrust that may clearly be greater than the one measured before the stoppage. The creep of the soil leads to the tightening of the ground around the microtunnel causing significant excess friction (see paragraph 3.1.3.3).

Strong deviations in trajectory can also induce much greater frictional forces than those expected. In fact, the frictional forces on the boring machine and the first pipes during trajectory corrections are clearly higher; the curvature of the pipeline causes an increase in the contact pressure between soil and pipe, which can lead to an occasional increase in the total thrust (see paragraph 3.1.4.3).

3.2.1.3. *Abrasiveness of the soil*

Abrasive soil can cause excessive wear of the cutting head (sites at Solaize (1990) and Dreux (1993), FSTT RS 4 report). And yet, the special feature of boring machines is that one cannot change the cutting tools during jacking. Thus, the cutting tools can get excessively worn out, and then break under the effect of stresses applied during excavation. As the tools get worn out or break, the excavation of the ground becomes more and more difficult, which then requires an increase in the thrust to maintain the penetration speed.

Furthermore, the diameter of the cutting wheel can also be reduced under the effect of abrasion, thereby significantly reducing the overcut around the machine, which leads to an increase in frictional forces.

The values of abrasiveness and hardness of the soil therefore need to be estimated and taken into account in order to determine the type and the service life of the considered tools and possibly the length of sections.

3.2.1.4. *Sticking of clay*

Clay can be responsible for blocking the wheel and difficulties in the extrusion of excavation material. In fact, clay mixed with water becomes sticky and leads to certain difficulties such as:

– agglomeration of excavation material and formation of lumps in the mucking system,

– clogging of the mucking inlet sieve,

– clogging of the cutting wheel.

These difficulties can sometimes lead to stresses that exceed the capacities of the machine, and at times lead to the immobilization of the boring machine. Suitable measures, however, help limit these risks (see Chapter 5).

3.2.2. *Damaged pipes*

Pipes for microtunneling are subjected to circumferential (weight of soil, overloads) and longitudinal (thrust drive) actions during installation. It is the drive thrusts, inducing the compressive stresses within pipes, which are largely predominant.

Moreover, it must be noted that the pipes, generally 2 m long, constitute an "articulated body" and can get more or less out of line depending on the deviations and alignment corrections imposed by the operator. Thus, the drive thrust is no more centered on the axis of the pipe and the distribution of load is no more uniform, which leads to a concentration of stresses. Beyond a certain angulation value between pipes, only one portion of the transverse section transmits the axial stress. The maximum stresses increase that much more as the contact surface between pipes reduces (see Figure 3.18).

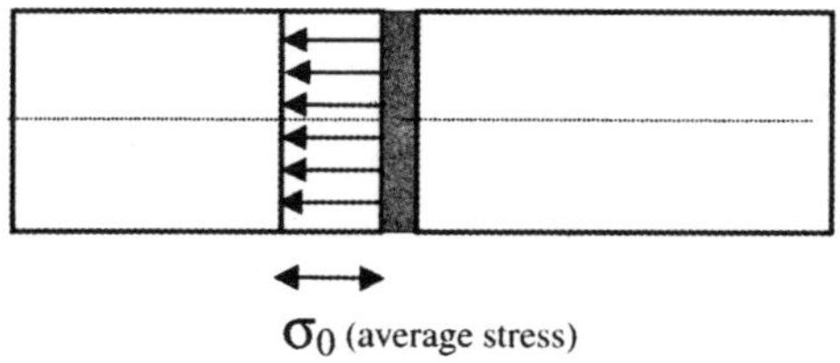

a) Case of two aligned pipes, centered jacking thrust

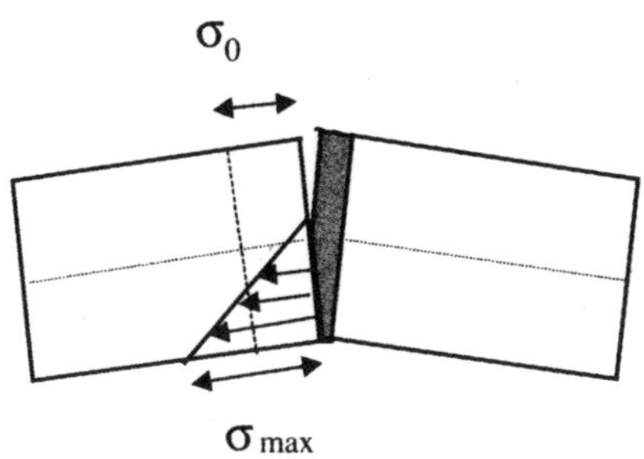

b) Case of two non-aligned pipes, off-centered jacking thurst

Figure 3.18. *Distribution of stresses in pipes*

In addition, the misalignment of pipes, which induces the offsetting of drive thrusts, can subject them to loads during bending that can even cause rupture in extreme cases.

The risk of punching of pipes by large blocks must also be noted. We were thus able to observe holes of 20–30 cm in the concrete pipes following jacking in formations containing blocks.

To calculate the permissible thrust that must not be exceeded at a site, it is essential to take into account the offsetting of loads. Section 6.4 shows different methods for estimating the permissible jacking thrust.

3.2.3. *Surface disturbances*

With microtunnels being constructed more often than not in urban areas at a low depth, it is important to provide for and control the possible disturbances at the surface that may result from digging by the boring machine. These disturbances can have many sources:

– settlements caused by closing of the annular space,

– settlements, upheavals or subsidences caused by an improper balancing of pressures at the face,

– upheavals of the ground caused by an excessively high injection pressure of the lubricant,

– resurfacing of the mucking fluid to the surface or in entrenched networks.

3.2.3.1. *Settlements caused by the annular space*

The excavation diameter is always greater than the diameter of the pipeline installed in order to facilitate guidance and reduce the frictional forces. This overcut therefore leaves an annular space between the ground and the pipes. With the soil mass progressively loosening, the annular space closes up, which could lead to compaction at the surface.

Very few experiments on actual structures have been carried out to date. Several authors have, however, successfully compared the results obtained experimentally with the geometrical parameters of the compaction basin calculated using the O'Reilly and New (1982) and Peck (1969) formulae.

This theory, which likens the profile of settlements at the surface to a Gaussian distribution, was established for tunnels of large diameters. Compaction profiles similar to the theory were demonstrated, and the maximum amplitudes of compaction measured at experimental sites turned out to be quite close to those estimated by the models of O'Reilly and New (Staheli *et al.*, 1996, Marshall *et al.*, 1996, Milligan *et al.*, 1995 and Rogers *et al.*, 1991). The amount of experimental data, however, remains insufficient to conclude on the systematic validity of these calculation methods for microtunnels.

3.2.3.2. *Instability of the face, poor balancing of the pressure at the face*

During jacking, the boring machine exerts a contact pressure Rp on the ground, which results from the force exerted directly by the cutting wheel and the pressure of the mucking liquid in the stope. This thrust is generally not measured, but Pellet (1997) showed that apart from the restarting phases, this thrust could be estimated as being the difference between the total instantaneous thrust and the envelope of the maximums of this total thrust.

The value of the pressure at the working face determines the equilibrium of the face:

– if the pressure is too low, there can be decompression at the face that can, in extreme cases, result in ground subsidence;

– if it is too high, the head of the boring machine will pierce the face and may cause, by pushing back the soil, significant movement in the soil mass as well as upheavals at the surface.

3.2.3.2.1. Decompression of the working face

Because of the small diameter, we sometimes consider that the pressure of the cutting wheel is sufficient to ensure the stability of the face, which leads to mucking with clear water. While the stability of the face can be effectively ensured in this way in soil with sufficient cohesion, it is not so in the case of soil that has low cohesion or is not at all cohesive. In this case, it is possible that in the absence of effective support (by pressure of the mud or pressure of the earth), the soil has a tendency to collapse at the face and the boring machine advances by causing an over-excavation. This can result in cavities, or even subsidence opening at the surface.

3.2.3.2.2. Blow-out of the soil

Disturbances caused by the upheaval of the ground at the surface as well as by the damage of a wastewater pipeline have been described by Phelipot (FSTT RS 22). They concern a site built in a clay pan of low compactness and resulting from the low resistance as well as sticky nature of this ground. Great difficulties in mucking caused the blocking of the attack chamber. The low resistance of this soil horizon, however, helped sink the pipeline by exerting a strong thrust on the face; this caused a partial blow-out of the ground, highlighted by the fissuring of the surface covering and by leaks from a wastewater pipeline damaged by excessive movement.

3.2.4. *Excessive roll*

Roll is the rotation of the entire boring machine with respect to its longitudinal axis. To limit the roll, the trailing tube of the machine is usually equipped with "fins" that help stabilize the machine body. A large torque or premature lubrication may favor roll. A large roll can cause the rupture of cable connections on the machine.

This roll can be created by the sudden stopping of the cutting wheel on a block in the ground; the rotational inertia then drives the entire machine and the tubes. These disturbances can have an impact on the quality of the joints between pipes, which can thus create defects of watertightness as well as difficulties in guidance: in fact, in phases of large roll, the laser beam may miss intercepting the guidance target. The operator must therefore provide a counter-roll to reposition the boring machine in its normal configuration, which can prove to be tricky in certain soil horizons.

Chapter 4

Guidelines for Investigations

4.1. General approach of the investigations

4.1.1. *General objectives*

Geotechnical investigations prior to the microtunnels project are considered here in a broad sense: they include research of geological, hydrogeological and geotechnical data, as well as the natural or artificial hidden obstacles, which can interfere with the course. They have four objectives:

– to contribute to the optimization of the geometry of the project (course plan, longitudinal profile, number and installation of shafts), knowing that many more constraints will have to be taken into account;

– to optimize the choice of the boring machine and dimension of the tubes to be jacked;

– to size the shafts and the "dead man";

– to outline an implementation method and allow the project manager and the companies to estimate the cost and construction time.

These investigations have to be carried out at the earliest stages of the project because it is then that the chances of saving money are the greatest, in view of the first objective in particular. The risk of carrying out investigations "in vain", which are outside the final course, is not as important compared to the savings that can be made during the design stage of the project.

It must be noted that the shafts generally represent between 20 and 40% of the cost of the microtunnel project and this cost increases rapidly if they have to be made watertight. An implicit objective of the investigations is to check that the layout designed has no insurmountable obstacles which could make the project unfeasible: this too must be done as early as possible.

The aim of this entire work is to establish a *mapping of the geological, hydrogeological and geotechnical hazards* as realistically as possible so as to better integrate the construction project with the risk level results measured or even interpreted: the mapping of the hazards should be the culmination of the first phase of the geotechnical knowledge regarding the soil and the sub-soil.

4.1.2. *Progress of the investigations*

The normal progress of the investigations includes four successive stages, which will be detailed in the following chapters:

– documentary survey,

– geophysical investigation,

– geotechnical boreholes and tests (*in situ* and in the laboratory),

– summary of all the investigations, and preparation of a geotechnical file to be attached to the tender documents.

For carrying out these investigations and the evaluation of hazards, the usual guidelines relating to all geotechnical projects remain valid:

– first look to locate the project in the regional and local geological setting, in order to be sure of better understanding the configuration of layers; this is the case all the more so for hydrogeological conditions;

– provide for an investigation program consisting of, preferably, two stages separated by an intermediate summary, which will help "rectify the drive" in view of the initial results;

– reserve about 20% of the available budget for the investigations or tests not allocated in advance, in order to have resources to react immediately in face of unforeseen results;

– encourage the intending contractors to visit the contracting authority and observe the core samples and study the detailed results of the investigations not attached to the tender documents.

4.1.3. *Cost of investigations*

Unlike trench work where one can adapt to the encountered soil on a day-to-day basis, the work with boring machines requires particularly careful investigations (therefore relatively costly), due to several reasons:

– the conclusions drawn from investigations will be irreversible: once the machine has entered the ground, it cannot be modified or changed, at least until the section is complete;

– the economy of the project, for the company as well as for the contracting authority, is incompatible with a serious overestimation of the penetration speed, and all the more so with a blocking of the machine between two shafts, which requires, more often than not, the machine to be "recovered" by digging an intermediate shaft;

– the possible damage to networks that are not investigated can lead to risks for the neighborhood, not to mention the costs, penalties or durations for insurmountable restorations.

For a microtunnel project, it is normal that the cost of investigations is proportionately greater than for a man-tended tunnel, and all the more so as the length reduces: the investigations/work ratio is currently about 5 to 15%.

This initial investment, designed to reduce the burden of additional costs, must be regarded as an integral part of the project and must be consequently planned and budgeted for. Thus, for a 200 m microtunnel, an investigation budget of the order of €20,000 is not too excessive.

These Figures can be compared to the guidelines of the NASTT, which recommends an average 0.5 m of boreholes per linear meter of the tunnel. These criteria for length must, of course, be weighted according to the geological variability and uncertainties peculiar to the considered site.

Of course, it is up to the contracting authority to bear the cost of these initial investigations: only he has the resources and above all the sufficient anticipation time to correctly carry them out.

Furthermore, in case of insufficient investigations, it is he who will bear the consequences in terms of the cost and time; similarly, it is he who will profit from the information obtained thanks to a good investigation.

4.2. Data to be acquired

4.2.1. *Geological configuration of the site*

Understanding the geological configuration of the site, which results in a longitudinal geological cross-section, is the fundamental objective, which is at the basis of the rest of this approach. This understanding can result only from *local data set back in a regional environment*, this latter enabling us to interpret it and then extrapolate it conclusively.

It is only after having delimited and prioritized these geological units that we can try and characterize each of them from the physical and mechanical point of view.

As an example, only a good knowledge of the regional geology will enable us to know if we risk encountering erratic blocks in the moraines, or large stones at the bottom of alluvia, or flint embedded in the chalk, etc., all things that boreholes have very little chance of detecting directly. Similarly, when boreholes encounter a substratum with variable dimensions, the design of the roof (and the estimation of the uncertainty associated with this design) can be conclusively done only on the basis of a known regional model of erosion or weathering.

The investigations must also endeavor to detect and characterize a certain number of objects or geological configurations that are particularly penalizing for boring machines, notably:

– the facies whose constituents are *mechanically heterogeneous* (anthropic backfills, moraines, burrstone clay, slope scree, irregularly altered rocks, flint embedded in a softer matrix, etc.);

– soils with *highly coarse sieve*, such as torrential alluvia;

– *plastic clay*, which can cause formidable problems of swelling or sticking in the presence of water;

– *soft soil*, in general, which causes guidance problems;

– *the interfaces between layers*, as they are generally much more difficult to overcome with a boring machine than enclosed grounds on their own; we will moreover endeavor to support the profile lengthwise in order to prevent such a situation, all the more so as it is generally impossible to locate in advance the interfaces to the nearest decimeter;

– finally, *the natural cavities*, either of karstic (in gypsum or limestone) origin or caused by extraction of fine materials in the non-saturated area.

4.2.2. *Hydrogeological conditions*

It is essential to know the probable and maximum level of the water table during the works, for the sizing of shafts as well as for the penetration of the machine. The knowledge of the permeability of the soil embedded in the water table is also necessary in order to develop a method for the execution and waterproofing of the shafts. Two secondary parameters may also be useful:

– the *chemical composition* of the underground water as well as its pH (particularly when it is rich in sulphates), for the possible interactions with the drilling fluid, for problems of sticking as well as for its aggressiveness in relation to the pipes to be jacked;

– the *horizontal speed of flow* of the water table, which, if very high, can hinder the lubrication of pipes during jacking;

– the *predictable fluctuations of the level of the water table*, which can vary between the time of the prior study and that of the works, and thus perhaps modify the project or the conditions of operation (hydrogeological risks in relation to the structure).

4.2.3. *Geotechnical characteristics of the ground*

The parameters useful for the development of a microtunnel project are indicated in Table 4.1, where we have distinguished between the mandatory parameters to be measured in all cases, and the additional parameters, which are generally to be estimated indirectly.

	Mandatory parameters	**Additional parameters**
Physical characteristics	Sieve Plasticity, Blue value Specific weight, water content	Mineralogy Ability to stick, pH
Mechanical characteristics	Resistance (during compression or short-term cohesion), deformability Abrasiveness and hardness	Creep and swelling parameters Dilatancy
Hydrogeological characteristics	Fluctuations of the water table level (minimum, average and maximum level)	Permeability Flow velocity

Table 4.1. *Useful geotechnical parameters for microtunnel projects*

4.2.4. *Cavities and artificial obstacles*

This aspect of the investigations is that much more important as the microtunnel needs to cross callows (depth < 5 m), and the site has been urbanized long before. The following obstacles must be investigated, as they are the most challenging for a boring machine:

– demolition and landfill products (concrete, scrap metal, etc.);

– old foundations (masonry work, piles, etc.) of buildings that have been subsequently demolished;

– old wells (originally there was one in the courtyard of every house);

– forgotten cellars, quarries or underground shelters that are often not filled in;

– various pipelines and cables, in operation or abandoned; these networks are more often than not located in the 0–3 m section, but their position is never completely in keeping with that indicated on the drawings, if these exist …

4.2.5. *Environmental conditions*

The importance of these parameters related to the geotechnical conditions is extremely variable depending on the situations. In all cases, it is necessary to examine at least the following points:

– *initial condition of the pollution of the ground*, concerning possible interactions with the drilling fluid and its additives, as well as the constraints that may result from this pollution by the stockpiling of earth (evacuation to a storage area considerably increases the cost of the structure);

– *evacuation of earth*: identification of possible locations for stockpiling, considering a possible initial pollution and the mucking products used;

– *condition of the existing frame* and neighboring underground structures, from the point of view of their sensitivity to possible movements caused by the boring machine (settlements and uplifting), as well as the permanent overloads that they transmit to the ground to be crossed;

– *construction of access shafts*: the locations and surfaces available, accessibility, sensitivity of the site to pollution (noise, soil, etc.).

4.3. Methodology and means of investigation

The means needed to be implemented to acquire the data listed in section 4.2 include work at the office, site and laboratory.

4.3.1. *Documentary survey*

The consultation of previous documents is essential, particularly for urban areas: there is always data of geotechnical interest to be found (drilling databases, geological maps, files of earlier structures, old drawings, etc.). The issue is all the more important as new investigations risk being difficult to carry out. This consultation will also provide additional indications that are indispensable for two reasons:

– the presence of existing or forgotten *networks and underground obstacles*; the first are noted (more or less accurately) on drawings of statutory companies or in databases of some towns;

– the historic levels attained by the *water table* in the past; we must note, in fact, that piezometry measured during a drilling campaign is subject to all sorts of influences (rain, rise in water level, blockage in neighboring catchment points, etc.), and will not be necessarily identical during jacking work.

We include in this stage the visit to the site, excavations or earth situated in a geological context that is identical or similar, considering that a "good" outcrop is more important than several drilling operations, even if it is not exactly on the course.

4.3.2. *Geophysical investigations*

4.3.2.1. *Objectives*

The use of geophysical methods has many advantages for the investigation of microtunnel projects:

– providing a 2D image on the distribution of the soil along the projected line route (most methods providing continuous profiles);

– optimizing the layout of exploratory boring (with or without core samples);

– the possibility of laterally extrapolating the boring data, once the geophysical profiles have been tested on these borings;

– finally, emphasizing the localized heterogeneities, which are not likely to be encountered by boreholes (which assumes the use of a geophysical method that is well suited to the nature of the "objects" sought and their host).

These various applications, quite considerable in number, show that it is preferable to *begin geophysical investigations* before boring.

4.3.2.2. *Usefulness of different methods*

Table 4.2 summarizes the main geophysical methods used during the investigation of microtunneling projects, with their advantages and drawbacks; only the most common methods have been listed, without prejudging new methods or combinations of methods currently being developed.

Method **Basic principle**	**Areas of application** **Adv. = main advantages**	**Drawbacks**
Geological radar Reflection of electromagnetic waves from the interfaces	Detection of interfaces, various networks and obstacles (metallic or non-metallic) Adv. Rapid, Not very cumbersome Continuous profile at high resolution	Tricky implementation and interpretation (by specialists) Blind in clayey soil or in the water table. Max. depth 5-10 m Penetration < 2m if networks Ø20 mm
RMT (radio-magnetotellurgy) Measurement of the resistivity via perturbations in the electromagnetic field of a radio transmitter	Geological identification of soil and buried obstacles (metallic or non-metallic) Adv. Continuous profiles, very rapid, cheap, Good lateral resolution. No geological negative guidance	Investigated depth not controlled well Disrupted by metallic networks, but this can be an advantage! Not very suitable for urban areas (strongly interfered signals)
Electromagnetic methods with close transmitter Creation of Foucault currents, measurement of the induced field	Geological identification of the soil Detection and location of metallic networks Adv. Easy to implement and efficient	Penetration depth: – EM 31 : 3-4 m – EM 34 : 10 m Discontinuous profiles, Frequent interference in urban areas
Electrical method Measurement of apparent resistivity by injection of direct current and measurement of the potential difference	Geological identification of the soil Adv. Control of the depth and the lateral and vertical resolution Suitable for all types of soils	Discontinuous profiles. In urban areas, ensure good electrical contact with the soil
Electrostatic quadripole Same as electrical method (current 10 to 50 kHz)	same as electrical method, but: – much quicker, continuous profiles, – can be used for road surfaces	Limited experience
Electrical method in aquatic sites Measuring the resistivity by current injection; measuring ΔP by electrodes dragged at the bottom of the water	Geological identification of the soil Adv. Control of the depth and the lateral and vertical resolution. Continuous profiles. Low cost, suitable for all types of soils	

Seismic refraction Refraction of seismic waves on layers at speed increasing with depth	Looking for the dimension of a substratum Assessment of the mechanical characteristics of layers	Assumes a low dip Poor horizontal resolution In town: frequent static requires lightweight sources
Seismic reflection Reflection of seismic waves on contrasting interfaces	Definition of the geometry of layers, in soil or in aquatic site Adv. Rapid method, providing a continuous profile	In town, preferable to work at night Under water, blind if the bottom is muddy with gas bubbles
Seismic surface waves (SASW) Analysis of the dispersion of seismic surface waves	Identification of hard or loosened spots (based on the distribution of shear modulus)	Assumes stratified ground Can be used in town (light sources) Not widespread in France
Microgravimetry Local variations of the gravitational field	Localized search for spaces, decompressed areas and rippling of the substratum	Poor resolution; small cavities (Ø < 2 m) not visible if depth > 5 m Low efficiency. Requires complex corrections in urban areas

Table 4.2. *Areas of use and limitations of geophysical methods*

The most widespread method is the *geological radar*. Its main purpose is to detect and locate objects of known nature but whose position is unknown (typically, networks and horizontal or vertical interfaces), rather than to identify discontinuities or unknown objects. It is also used to "indicate" the presence of heterogeneities, with an excellent resolution (5 to 10 cm at less than 5 m depth); but their exact nature will have to be specified by other means of exploration.

Unfortunately, the radar is not very effective in silt-laden soils, and blind in clayey soils or in the water table. Very precise calibration tests were carried out in 1997 at the experimental LCPC site at Nantes, which helped prepare a "catalogue of radar signatures" for the most common obstacles (FSTT RS 17).

After the radar, several geophysical methods with large efficiency must be listed:

– *radio-magnetotelluric* (RMT), a method that is suitable especially for unused sites and which provides continuous profiles, such as the radar;

– *electrical prospecting by electrostatic quadripole* (also continuous profiles);

– *electromagnetic prospection* (type EM 31 or EM 34).

These very rapid methods are better suited than the radar to describe the distribution of soils and their nature, but they are less effective in detecting and especially precisely locating obstacles and horizontal interfaces.

The seismic methods also need to be mentioned, in spite of their difficulties for use in urban areas, in particular:

– *seismic refraction*, which gives very good results in geological configurations characterized by a series of layers of increasing velocity with the depth, even under the water expanses of low depth;

– *high resolution seismic reflection*, which helps point out the possible existence of "reflectors" linked to contrasts in density and/or rigidity of layers, and which is very well adapted for the study of the crossing of water courses; in the case of shallow water courses (< 10 m), it is recommended instead that seismic refraction or *electrical prospection by direct current* be used.

Finally, *microgravimetry* needs to be mentioned, a localized method used to detect the presence of cavities whatever the nature of the host ground; but this method is effective only for cavities with low depth (a ∅ 2 m gallery remains invisible if its roof is higher than 5 m); it is quite expensive as it requires measuring stations that are very close together, particularly if one is looking for small cavities.

4.3.2.3. *General guidelines*

In all cases, one must first endeavor to design a projected geological section based on the documentation, estimate its uncertainty and then consult specialists to select the most suitable geophysical method(s). For a microtunnel project situated at a depth of less than 5 m at the town centre, where the geological structure is largely known, it is recommended to start with the following geophysical methods:

– geological radar, except in the case of clayey soils or the surface water table,

– in other cases, electrical or electromagnetic prospection EM 31 (to be replaced by EM 34 for a project located at a depth of between 5 and 10 m).

For a project in a ground that is less obstructed (outside the town centre), we will first try and clarify the geological structure over the entire course planned by using RMT, or seismic methods if the ground is suited for these. The detection of isolated obstacles will be done subsequently, by geological radar or electromagnetism. Microgravimetry will be used more frequently only to precisely locate the surface cavities whose existence is proven, but whose exact position is not known.

Finally, it must be indicated that the main French geophysical operators have signed a charter called the "Code of good practices in geophysics", which must be respected in the specifications.

4.3.3. In situ *boreholes and geotechnical tests*

4.3.3.1. *Objectives of boreholes*

Exploratory boring is indispensable as it meets several needs:

– it precisely defines the geological section of the course,

– it measures the level of the water table and possibly the permeability of the ground,

– it takes samples for geotechnical tests at the laboratory,

– eventually carrying out geotechnical tests *in situ*.

In the general case, where there are no observable outcrops, at least one part of the boring must enable a *direct visual description* of the nature of the ground and the taking of a sample. Only core boreholes and holes with a mechanical shovel, or certainly the SPT and augers fulfill this objective (the latter constitutes a good complement to enable an interpolation between core boreholes).

4.3.3.2. *Layout of boreholes*

The boreholes are drilled according to the results of the geophysical investigations, as they contribute to its interpretation just as geophysics helps extrapolate their results. The number of desirable boreholes depends strongly on the geological complexity of the site and the degree of its prior knowledge. As a guide, we can consider that at least one point of information is necessary for every shaft, with an average distance between boreholes of the order of 30 to 50 m for each section.

Urban cluttering very often hampers the layout of boreholes; however, it is important to note that it is not always essential to drill boreholes exactly on the planned course:

– first because they have a very remote chance of encountering local heterogeneities,

– then, because information outlining the reality is very often more precious: for example, it enables us to measure the transverse dip of the layers, or sometimes even improve the horizontal alignment.

Similarly, it is important to extend boreholes several meters under the maximum depth that can be envisaged for the microtunnel, in order to better understand the configuration of layers and make the interpolations between boreholes more reliable.

4.3.3.3. *Types of* in situ *tests*

The principal types of *in situ* tests that can be used are as follows:

– SPT (*Standard Penetration Test*), which provides reworked samples as well as an index N of resistance to penetration; but this index does not differentiate between the cone stress and the lateral friction;

– the static penetrometer or CPT (*Cone Penetration Test*), which enables, on the one hand, the measurement of the cone stress and the lateral friction separately, and on the other hand, provides a continuous resistance log, which can be correlated with the stratigraphy;

– the dynamic penetrometer, which provides a primary resistance index, quite continuous but difficult to interpret;

– the pressure meter, which helps identify the ground (if it is drilled with an auger) as well as provide deformability and resistance values.

4.3.3.4. *Guidelines on the choice of boreholes and tests*

Table 4.3 gives indications on the adaptation of various methods of geotechnical investigation for five types of frequently encountered soils – considering that each site is a specific case and that the level of previous knowledge of formations crossed is bound to have an influence on the investigation strategy.

Ground	Mechanical shovel	Core sampling	Destructive drilling	Auger	SPT	CPT	Dynamic penetrometer	Pressure meter	Notes
Urban backfills	**	*	**	0	0	0	*	*	Depends on the nature of remains
Fine soil (silt, clay)	*	**	*	**	**	**	*	**	
Sand and gravel	**	0	*	**	*	*	*	**	Depends on the Ø max of pebbles
Soft rocks	0	**	**	*	*	0	*	0	Depends on the depth of overs
Rocks not altered much	00	**	**	00	0	00	00	00	Depends on the degree of weathering

*Key: **: very well suited 0: not suited *: quite suited 00: unsuitable*

Table 4.3. *Adaptation of various geotechnical investigation methods to some typical soils*

NOTE: some particular types of ground require special attention, in particular:

– marl and limestone marl, which are intermediates between fine soil, due to their clay content, and soft rocks, due to their resistance;

– heterogeneous soils, such as blocks embedded in a more or less consistent clayey matrix (for example, grinding grit clay in the Paris region), where the methods used will give priority to shoveling pits, coring and destructive drilling.

Some general guidelines complement Table 4.3:

– in theory, the investigation tool that comes closest to the actual operation of a boring machine is the static penetrometer (CPT); its use is recommended when the frictional characteristics of the ground are essential (for large distances between shafts, in particular);

– the three types of penetrometers provide a cheap method for verifying, between boreholes with cores or auger, the consistency of the geology and the resistance of the soil, when there is a danger of detrimental variations;

– to size the shafts and *dead man* in loose soil, the pressure meter is the most suitable tool; in cohesive ground, core drilling can also be carried out with collection of intact samples for mechanical tests at the laboratory.

4.3.4. *Geotechnical tests at the laboratory*

The physical and mechanical properties of the ground used for a microtunnel project are listed in Table 4.1; they are measured by geotechnical tests carried out in specialized laboratories, or by *in situ* tests, as indicated in Table 4.4.

It is vital to emphasize the essential importance of *physical parameters*, which largely determine the conditions for excavation, penetration and mucking; they can be measured on lightly reworked samples, which can be collected using the auger or by SPT.

On the other hand, the measurement of mechanical characteristics at the laboratory requires intact samples from core boreholes or the mechanical shovel.

4.4. Contents of the geological record

The obvious interest of the contracting authority and its project manager is to make the investigation results available to the bidding companies in a form that is as clear and as complete as possible. In this way, the companies will reduce, as far as possible, the margins for vagaries that they will naturally include in their estimate.

The geological record attached to the tender document must generally include two kinds of documents:

– on the one hand, *the factual results of investigations*, which often include large amounts of data, not necessarily validated or properly correlated amongst one another;

– on the other hand, the "*Memorandum of geotechnical summary*", in which the contracting authority clearly summarizes the geotechnical conditions of the project while indicating and defining the residual uncertainty. The purpose of this memorandum must not be to make the company discard the geological hazards by vague or ambiguous drafting, but to enable it to correctly handle the unexpected, without litigation or stoppage of work at the site.

Parameters to be measured Tests	Techniques employed	Application of parameters						
		Shafts	Dead man	Excavation	Mucking	Thrust	Coating	Others
Laboratory tests								
Grading curve Specific gravity (γ_h and γ_d) Water content (w)	Sieving. Sedimentation analysis Heat curing, weighing Heat curing, weighing	x	x	x x x	x x	x x	x x	
Liquid limit (w_L), Plastic limit (w_P), plasticity index (IP) blue value	Atterberg limits Methylene blue test			x x	x x	x x	x x	Sticking, swelling
Simple compressive strength (σ_c) Un-drained cohesion (c_u) Cohesion (c'), angle of friction □	Crushing with the press Triaxial tests UU Triaxial tests CD or CU with measurement of u	x x	x x	x x x		x x	x x	
Abrasiveness	CERCHAR or LCPC Abrasiveness test			x				
Aggressiveness or pollution of the soil	Chemical analyzes				x		x	
***in situ* investigations**								
Geological nature of the ground	Boreholes with core or shovel	x	x	x	x	x	x	Samples for laboratory
Pressiometric module (E_M limit pressure(p_l))	Pressiometric test	x	x			x		
Number of strokes SPT (N) Cone resistance, lateral friction	SPT test CPT test	x x	x x			xx		
Permeability	Surge or injection testing	x		x				
Piezometric level	Piezometric measurement	x	x	x	x	x	x	x

Table 4.4. *Parameters and geotechnical tests useful for microtunnel projects*

From the contractual point of view, this memorandum summary is the main document to which the factual documents are subordinated; all that is asserted in it is thus the responsibility of contracting authority and it must therefore be drafted very carefully. It must include the following elements at the least:

a) the list of information sources used (previous reports, surveys, geophysical surveys, etc.), with a layout plan for boreholes and geophysical surveys in relation to the projected structure;

b) a precise geological description of the ground encountered;

c) a geological profile of the course with boring reports and other sources of information, in such a manner that the uncertainty inherent in the interpolations between boreholes is clearly highlighted; this longitudinal profile may, in some cases, be supplemented by cross profiles (when a large transverse variability is expected);

d) a table of values that are characteristic of every geotechnical parameter to be taken into account for the sizing of structures;

e) a list of hazards that are likely to be encountered and which the contracting authority wishes to indicate in advance, by estimating their probability as well as possible even if they cannot be located with precision.

Finally, the tender document could contain evaluation elements relating to the impact of the nature of the ground on the design of the structure and on the methods of execution.

Chapter 5

Guidelines for the Choice of Machines and Attachments

5.1. General information

Chapter 2 presented the principles of operation of boring machines, with their following main functions:

– excavation of the ground and stabilization of the face,

– removal of earth (mucking),

– monitoring and correction of the trajectory,

– installation of pipelines by jacking.

The last two aspects, which vary very little between different machines, are hardly determining factors for their choice. On the other hand, the experience of sites shows that certain difficulties encountered, particularly in relation to the ground to be excavated and evacuated, could have been limited or even avoided with equipment that was better suited for the ground. This of course assumes having a good prior knowledge of these grounds, their nature (clay, sand, rock), their condition (humidity, plasticity, compactness), and above all their heterogeneities to which these sites are very often highly sensitive; these geotechnical investigations form part of Chapter 4.

But even when we know the ground well, the choice of machines and their attachments is often tricky:

– firstly because the small dimension of structures makes the digging very sensitive to variations in the nature of the ground, even at the decimetric scale, as the excavation section can vary very rapidly from rocky soil (blocks embedded in a matrix), or even granular soil to a clayey soil (alternating layers of sand, silt and clay);

– then because it is not possible to intervene on the machine, to change the attachments (cutting tools) for example, before having reached the exit shaft of the section;

– finally, because in case of a major incident (blocking on an obstacle or following excessive friction caused by bulky clayey ground, for example), the recovery of the machine requires jacking from the surface or a traditionally drilled tunnel from the exit shaft, operations that are all very costly and time-consuming with respect to the initial project.

Thus, following the example of large tunnel borers which currently are the subject of technological developments aimed at designing "universal" machines, i.e. machines capable of digging in all types of ground, the problems for boring machines are comparable or even amplified because of their specific nature.

The object of this chapter is to provide some recommendations or guidelines that need to be followed in choosing machines and their attachments, depending on the current state of knowledge and technology. Of course, the machines are likely to evolve in the future, and therefore provide other answers to the questions raised here.

5.2. The choice of machines according to their mucking process

During the 1990s, we witnessed the development of three types of machines mentioned in Chapter 2:

– *machines with hydraulic mucking*: it turned out that these machines were well-suited in granular soil (sand and gravel) and in silt, whereas plastic clayey soil presented risks of sticking, leading to the clogging of the stope and/or mucking pipelines. In fact, this last aspect is very directly dependent on the choice of the mucking fluid (water or bentonite slurry) and the additives used, which will be tackled in Chapter 6. In addition, this type of machine requires the implementation of a plant to separate the earth, which represents a drawback;

– *machines with mucking by Archimedean screw*: these machines helped successfully build sites in ground with sand as well as clayey silt, but with limitations in ground with gravel (difficulties in evacuating the earth by the screw for the coarsest elements) and in plastic clay (sticking of clay galls in the screw). On the other hand, they did not require the installation of a plant for treating the earth, as it can be evacuated in the same state as it comes out of the machine;

– *machines with pneumatic mucking*: there are very few applications in France, and generally in sandy-gravel ground, for which these machines seem to be well-suited. On the other hand, the effectiveness is significantly reduced in cohesive ground (compact clay), and this is despite the possibility of injecting water under high pressure (or compressed air) directly at the level of the tools, to cut the ground.

Table 5.1 thus suggests the fields of application of these different types of machines according to the principal constituent of the ground.

Machines for mucking	**Clay**	**Pebbles**	**Gravel**	**Sand**	**Silt**	
					Not very plastic (IP < 30)	Plastic (IP > 30)
Hydraulic	**	**	**	**	**	*
With screw	O	*	**	**	*	O
Pneumatic	O	**	**	**	*	*

**: machine well suited; *: machine that can be convenient; O: machine not recommended

Table 5.1. *Choice of machines according to the type of ground to be excavated*

However, since the end of the 1990s, there has been a virtual disappearance of screw type or pneumatic mucking machines in France, and machines with hydraulic mucking are thus tending to become the "only tools"; the possible difficulties of sticking relating to plastic clayey ground have been remedied by an adaptation of bore fluids and their additives.

5.3. Choice of attachments

Whatever the type of machine, the attachments that make up the cutting wheel and the ground cutting tools, the crusher built into the head and various other attachments (particularly injection nozzles), are essential elements; in fact they determine the blasting of the ground and the subsequent possibility of evacuating the earth by the mucking circuit.

The main characteristics of the ground that determine the choice of these elements are:

– *the hardness*, in case of rocks (or anthropogenic obstacles), that exists over the entire section to be excavated: this is the case for microtunneling in rock or in blocks of dimensions greater then the boring diameter: in this case the blasting tools must enable the cutting up of the rock into elements of sufficiently small size so that these can enter the chamber, and be compatible with the mucking circuit. The "rock" type machines can drill through rocky ground whose compressive strength σ_c is 200 MPa;

– *the presence of hard isolated elements in a soil matrix*: it is then the intrinsic hardness of blocks, their dimensions and the compactness of the matrix containing these that are important. If fact, if the matrix is not very compact, the large elements that it contains risk getting pushed back by the wheel, sideways or forwards; or else they can rotate with it, but they will not penetrate into the chamber. The penetration will thus be greatly inhibited. We generally consider that the blocks of dimensions less than 1/3 of the boring diameter D do not pose major problems, provided the machine is equipped with a crusher;

– *susceptibility to sticking of clay*: in the case of sticky shale (i.e. clay that can be characterized by a plasticity index IP > 30 approximately), the chips cut by the head have a tendency to adhere, thereby forming a doughy mixture. This will at least result in an increase in the rotating torque of the head, or even the plugs in it and the mucking system;

– *abrasiveness*: in the case of highly abrasive ground (siliceous pebbles, flint, etc.) we could be induced to reinforce the tools (rotary cutters or teeth) with tungsten carbide discs in order to reduce their wear.

Table 5.2 constitutes a guide for difficulties that can result due to these geotechnical features, the functionalities sought to remedy them, and indications on the choice of corresponding equipment.

5.3.1. *The heads: opening, cutting tools*

The machine head is equipped with:

– a cutting wheel whose tools enable the blasting of the ground, under the combined action of rotation and thrust;

– a crushing cone, located behind the cutting wheel and designed to reduce the size of the largest elements to enable their mucking.

The cutting heads (see Figure 5.1) must be adapted to various types of ground. They differ from one another by their cutting tools and by the geometry of the wheel, particularly the size of openings enabling the passage of the ground. Based on the indications provided above, Table 5.2 proposes the main guidelines according to type of ground to be excavated.

Ground	Determining characteristics	Compactness of the matrix	Functionality sought	Choice of attachments
Rock full section	σ_c of blocks < 200 MPa	Not applicable	Cutting the rock to reduce it into elements of small dimensions	Disc cutters Openings of the head of dimensions Compatible with the mucking circuit
Blocks packed in a matrix	Dimension of blocks < D/3	Not applicable	Enable the blocks to penetrate into the head Reduce the blocks to dimensions < diameter of the mucking system	Open head Crusher
	.Dimension of blocks > D/3 and σ_c < 10 MPa	Loose	Cutting of blocks relatively easy Avoid pushing too much	Bits (+ disc cutters) Head very open Standard crusher
		Compact	Cutting of blocks relatively easy	Bits (+ rotary cutters) Head open Standard crusher
	Dimension of blocks > D/3 and σ_c > 10 MPa	Loose	Cutting of blocks very easy Avoid pushing too much	Disc cutters and Bits Head very open Powerful crusher
		Compact	Cutting of blocks difficult	Disc cutters and picks Head open Powerful crusher
Silt	IP < 30	–	–	Bits Head very open
	IP > 30	–	Scrapers	Clay Injection nozzle on the head (*jetting*) Head very open

Table 5.2. *Guide to the selection of excavation equipment*

1) For sandy-gravely ground | 2) For cohesive ground | 3) For rocks

Figure 5.1. *Different cutting heads (Herenknecht document)*

1) *For sandy or gravely soil* (particularly alluvial), the cutting wheels are equipped with teeth and are relatively open to facilitate the entry of the soil into the chamber where, if necessary, it will be crushed by a crusher to reduce the larger elements and allow them to penetrate into the mucking circuit. In ground containing coarse elements, these teeth expose the blocks, which are then crushed; if there is a risk of encountering blocks of large dimensions (> D/3) and strong resistance (σ_c > about 10 MPa), it will be necessary to also provide rotary cutters on the wheel in order to be able to cut up these blocks.

2) *For cohesive soil* (silt, clay, marl), the cutting wheels are equipped with tools ("blades" or picks) for cutting soil chips. The crusher is generally useless, unless the clay contains blocks. The head is usually wide open, in order to facilitate the entry of the soil into the chamber, particularly if it has a tendency to aggregate (sticky clay). To make the evacuation of soil easier in the case of sticky clay, on some machines, water at a higher pressure may be injected on the wheel and in the stope to prevent sticking of the clay and clogging of the mucking circuit.

3) *For rocks*, the cutting heads are equipped with rotary cutters and have small openings. In fact, the rotary cutters, by transmitting the thrust exerted on the machine, crush the rocks by applying shear and tensile stresses on them, which propagate cracks and loosen the chips. Such machines can drill in rocky ground whose resistance to compression is 200 MPa. This type of cutting wheel, also used in soil containing large blocks is, on the other hand, not suitable for loose ground.

5.3.2. *The overcut*

The overcut is the difference between the outer radius of pipes and that of the excavation made by the cutting wheel. Providing for a sufficient overcut, suitable to the nature of the ground, is an essential factor for the success of a microtunneling project. In fact, the value of the overcut has a predominant influence on the frictional forces along the pipes (see section 6.2):

– in stable ground, it enables a better distribution of the lubricating fluid around the pipes, which in turn improves the confinement of the excavation and thereby prevents the closing of the annular space under the effect of unloading;

– in dense sand, it helps compensate for the dilatance of the ground at the soil-pipe interface. In fact, in the absence of an overcut, the normal stress on the pipes increases under the effect of dilatance, and correlatively the frictional forces also increase;

– in swelling clay, it helps avoid the "tightening" of the pipes under the effect of swelling of the ground.

On the other hand, an excessive overcut can cause problems of roll for the machine (see paragraph 7.2.3). Such a roll can lead to settlements at the surface by closing of the annular space, and also to disturbances in the pipes subjected to differential thrusts due to their compression between the machine and the thrust frame.

The overcut is obtained by tools placed at the periphery of the cutting wheel, which jut out slightly from the latter. In France, and for boring machines of 600 to 1,200 mm diameter, an overcut of 20 to 30 mm is usually employed, and no problem of settlements at the surface has been observed; more than the overcut, it is the instability of the face that can cause improper control of the mucking and thereby result in ground movement.

5.3.3. *The crusher*

The crusher is a tapered element placed inside the chamber, which is also tapered but with a greater cutting angle. The blocks are thus progressively driven into spaces of smaller and smaller dimensions, and thereby crushed. The movement of the crusher can be:

– either circular: it is only the penetration of the machine which pushes the blocks between the crusher and the chamber,

– or epicyclical, with the help of an eccentric disk, which facilitates the crushing of blocks and pebbles.

5.3.4. *Bore fluids*

Bore fluids, which are used for mucking in the case of boring machines with hydraulic mucking or for lubrication all along the string of pipes, are not very important factors in the choice of the machine and its attachments at the head, as they have very little influence on the blasting.

On the other hand, they play a role during jacking that is sometimes a determining factor for the behavior at the soil-pipe interface all along the pipeline:

– in granular soil, by exerting a confining pressure on the contour of the excavation, thereby reducing the frictional forces,

– in clayey soil, by limiting the tightening stresses linked to the "swelling" of the clay.

In both cases, it is essential to remember that the role of the lubricant can only be fully effective if there is injection of fluid in the annular space at various points on the string of pipes, and not only from the head; this supposes that the fluid is injected with a continuous flow, to limit the settlements on the one hand, and to compensate for the loss by penetration of this fluid in highly permeable ground on the other. We will however note that boring with a small diameter (< 700 mm) greatly limits the possibilities of lubrication at several points.

In addition, in the case of swelling clay that is not saturated initially, we must ensure by a judicious choice of the fluid that the water content of the fluid is not in itself a factor that increases the swelling.

The microtunneling system will therefore have to provide for the possibility of injecting these lubrication fluids continuously and at several points. Indications on the types of fluids suitable for various grounds and their implementation will be given in sections 6.5 and 7.3.

Chapter 6

Guidelines for Project Design, Dimensions of Pipes and the Pipe Jacking System

6.1. Design of shafts

It should be recalled that the pipeline is jacked in a straight line between a starting shaft and an exit shaft, shafts that can, if necessary, be used at the final stage as clean-outs. The different functions of the starting shaft are as follows:

– installation of the spilling station, with its supporting downstream shell,

– installation of the laser measurement system,

– setting up of the boring machine and the pipe elements,

– removal of earth.

The exit shaft is used only for removing the boring machine.

The shafts constitute an important element of the project, particularly due to their cost, which often represents 20% to 40% of the total budget. The optimization of a microtunneling project is therefore limited by the number and dimensions of these shafts (in depth and in section), and this is an important part of the planning of a project:

– the number of shafts obviously depends on the nature of the project (total length, installation, possibility of clearing rights of way of the site at the surface) but also on the maximum jacking length planned for each section;

– their depth is evidently directly related to the longitudinal profile retained for the project: one must therefore keep in mind that, when the topography and operation of the project permit, it is more economical to design profiles with a low depth.

The dimension of the starting shaft depends on:

– the length of pipe elements (generally one to two meters, or even three meters);

– the overall dimension of the jacking frame;

– the working area required for various connections (cables, mucking pipelines, etc.);

– finally, the overall dimension of the thrust frame.

However, the dimensions of the exit shaft depend only on the length of the boring machine.

For 2 m long pipe elements, the dimension of the starting shaft is currently 4×4 m^2 or even 3×4 m^2 depending on the diameter of the pipes, the exit shafts being of smaller dimension (3×2 m^2). The shape can be rectangular, circular or oval; the choice essentially depends on the reinforcing techniques retained and the orientation of the drive, that is to say the horizontal alignment of the structure.

The exit and starting shafts require, in almost all cases, a reinforcement that acts as a retaining structure for the ground, as well as a waterproofing layer when the unconfined groundwater is reached. Several reinforcing techniques may be used (Schlosser and Leca, 1992); they do not differ from the traditional techniques used classically in heavy and highway constructions.

– *reinforcing with metallic sections*, with wooden, metallic or concrete planks inserted between the sections if necessary (dry ground);

– *shotcrete*, with or without reinforcement (except for the unconfined groundwater);

– *sheet pile wall* (applicable under the unconfined groundwater, but that can be difficult to drive in soil that is very hard or coarse);

– *cutting with nozzles or precast segments* in reinforced concrete: this technique helps construct permanent structures, outside water as well as under the unconfined groundwater;

– *diaphragm wall*, particularly well-adapted to deep shafts under the unconfined groundwater, but requiring the use of heavy equipment, at times out of proportion with the dimensions of the shaft. Slurry walls can often prove to be more economical;

– *enclosure with jet-grouting connecting columns*, requiring specific material (much lighter than those of diaphragm walls) and highly skilled personnel. Equally well-suited for loose soil and relatively deep shafts, it requires great precautions when the reinforcing must play the role of a waterproof curtain (shaft under the unconfined groundwater) in caving formations, due to the risk of windows between columns (in case of heterogeneities of the column diameter or drilling deviations).

The choice will have to be made according to the nature and properties of encountered soils, as well as the presence of unconfined groundwater at a depth that is less than that of the shaft, in which case the reinforcement will have to fulfill the function of a supporting as well as a waterproofing structure. In certain cases the processes for treating the ground such as injections or lowering of the unconfined groundwater may help avoid heavy reinforcements.

The retaining structures (apart from the thrust wall – see below) are also sized in a classical manner (calculation of passive earth pressure or by reaction modules) by taking into account the low dimension of the shaft, which often enables a significant reduction of stresses on the reinforcements.

The downstream shell of the jacking cylinders (also sometimes called "thrust wall"), installed in a shaft must be sized to be able to absorb, by supporting against the ground, the maximum jacking stresses. Their sizing may be done:

– either by a simple *stop calculation*, with the distribution proposed by Stein *et al.*, 1989 (see Figure 6.1): stress V in the jacks, distributed on the thrust wall of dimensions $h_2 \times b$, must not exceed the stop stress that can be mobilized in the ground. This method helps determine the permissible stress V_{adm} for spilling by the formula:

$$V_{adm} = \frac{K_p.\gamma.b}{2.F}\left[h_1^2 + h_1.(2h_2 + h_3)\right]$$

where K_p is the passive earth pressure coefficient, γ the specific weight of the ground, b the width of the *dead man*, F a safety coefficient that can be taken as 1.5, and h_1, h_2 and h_3 the geometrical parameters described in Figure 6.1.

Note: in certain cases of unsuitable ground, one must ensure that the downstream shell is loosened from the supporting structure of the shaft in order to avoid mobilizing the entire supporting structure, which could lead to disturbances in the latter.

– or by a *deformational calculation*, such as the calculation of the coefficient of subgrade proposed by SIA (a French company of engineers and architects) mentioned by Schlosser and Leca (1992). The abacus of Figure 6.2. directly gives the permissible forepoling stress V_a according to the dimensions of this *massif* and the modulus of subgrade reaction, noted E, according to two criteria: a resistance criterion (curves A: ϕ' constant) and a deformation criterion (curves B: $E/\gamma.h$ constant).

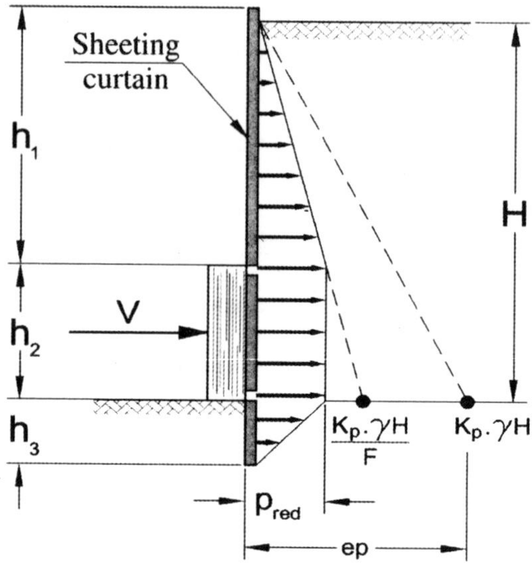

Figure 6.1. *Calculation principle of the "stopped" dead man*

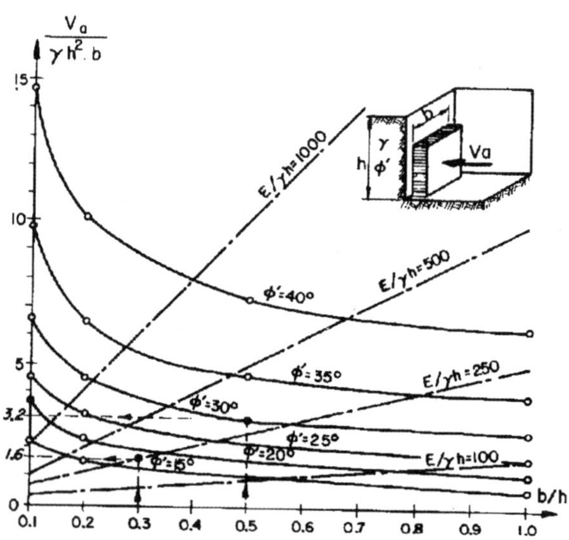

Figure 6.2. *Sizing abacus for the dead man in stresses and deformations*

6.2. Calculation of pipe jacking stresses

The thrust required for jacking P_{total} is determined by adding the pressure at the head of the boring machine R_p and the frictional forces F that are exerted on the pipeline in contact between the soil and the pipes (see Figure 6.3). The latter increases with the drilled length and generally constitutes the predominant part of jacking stresses at the end of shaft sinking.

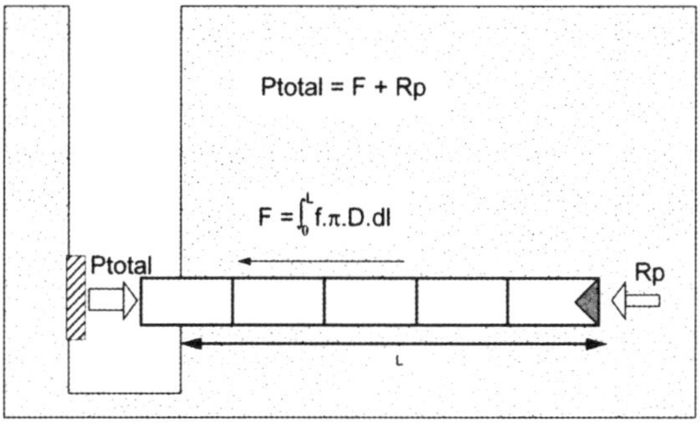

Figure 6.3. *Schematic diagram of pipe jacking stresses*

6.2.1. *Definition of friction between the soil and the pipes*

The friction between the soil and the pipes depends on the nature and condition of the soil, the nature and condition of the pipe surface as well as other parameters such as the depth, size of the overcut, lubrication or stoppage in jacking. This leads to distinguishing between several types of friction in relation with some of these parameters.

6.2.1.1. *General definition*

In general, we can define a local frictional stress f, commonly known as unit friction between the pipes and the soil, according the following relation:

$$f = \frac{dF}{\pi . D_{ext} . dL}$$

with D_{ext} the outer diameter of the jacking pipes, and dF the variation in frictional forces over a length dL.

This unit friction may be determined from the change in the thrust stresses recorded during microtunneling operations. It can also be defined from the effective stress σ ' normal to the pipes and the ground-pipe frictional coefficient μ :

$$f = \mu * \sigma'$$

6.2.1.2. *Specific friction values*

– *Dynamic friction f*: this is the friction during jacking, when the pipeline is advancing within the soil.

– *Static friction* f_{stat}: this is the friction that is caused after a stoppage in jacking. Due to the creep in most grounds, this friction is generally greater than the dynamic friction.

– *Lubricated friction* f_{lub}: in order to reduce the frictional forces, lubricating products are injected between the ground and the pipes, which leads to a value f_{lub} of the unit friction.

In addition, due to the difficulty of knowing the value of the stress at the head on the site, certain empirical results include the stresses at the head in the friction or adopt a conventional value for this stress. We may thus define:

– *The apparent average friction* f_{app}: it is equal to the total stress related to the surface of the pipes jacked, i.e.:

$$f_{app} = \frac{P_{total}}{\pi.D_{ext}.L}$$

– *The average conventional friction* f_{conv}:

$$f_{conv} = \frac{P_{total} - R_{pconv}}{\pi.D_{ext}.L}$$

where R_{pconv} is an arbitrary value assigned to the stress at the head.

6.2.2. *Experimental results relating to unit friction*

6.2.2.1. *Results of the French National Research Project "Microtunnels"*

During the French National Research Project "Microtunnels", 14 sections helped determine the unit friction experimentally (Pellet, 1997), (Phelipot, 2000). These 14 sections consist of a total length of 1590 m and a relative depth H/De of between 3.5 and 14.5 (average H/De = 8).

Tables 6.1 and 6.3 present the dynamic friction values during jacking, obtained with or without the injection of lubricating fluid in the annular space. Three major categories of soil were distinguished: fine sand (Table 6.1), coarse sandy-gravel soil (Table 6.2) and mainly clayey soil (Table 6.3).

	f (kPa)	f_{lub} (kPa)	Geological formation	Nature and characteristics of the ground
Sandy ground	5.2	1.9	Rubble	Fine sand sometimes gravelly, loose ($p_{l,moy}$ = 0.53 MPa)
		1.9	Fontainebleau Sand	Fine sand with clayey and compact levels ($p_{l,moy}$ = 2.4 MPa)
	7.3	1.7	Fontainebleau Sand	Fine clean sand
	5.6	1.6–2.7	Garonne Alluvia	Fine clean sand
	4.5	0.6–4.9	Garonne Alluvia	Fine clean sand
	4.5	0.5	Garonne Alluvia	Fine clean sand
Average	*5.4*	*2.0*		

Table 6.1. *Values of dynamic unit friction in sandy soil*

	f (kPa)	f_{lub} (kPa)	Geological formation	Nature and characteristics of the ground
Sandy-gravely soil		2	Old alluvia	Very compact, medium to coarse sand
		6.5	Old alluvia	Very compact sandy gravel and gravel
		8–10	Old alluvia	Very compact sandy gravel and gravel
	1.8 – 5.2		Altered Gneiss	Gneiss with semi rocky consistency ($p_{l,moy}$ = 3.5 MPa)
	5.2 – 17.2		Gneissic sand	Loose clay loam soil ($p_{l,moy}$ = 0.9 MPa)
	2.2		Altered Gneiss	Gneiss with semi rocky consistency ($p_{l,moy}$ = 3.5 MPa)
			Gneissic rock sand	Clean sand and gravel with small amounts of loose clay ($p_{l,moy}$ = 0.29 MPa)
	10.8		Anthropogenic backfill	Clean clayey gravel
Average	*7.4*	*6.9*		

Table 6.2. *Values of dynamic unit friction in sandy-gravely soil*

	f (kPa)	f_{lub} (kPa)	Geological formation	Nature and characteristics of the ground
Clayey ground	5.8	3.3	Brie limestone	Silty marl (IP = 36) and sandy marl (IP = 19)
	5.3	2.8	Beauchamps Sand	Lightly sandy marl (IP = 19) with compact clayey gravel (IP = 4.5) ($p_{l,moy}$ = 1.7 MPa)
	1.35		Alteration of vosgien clay	Clay-sandy silt (IP = 19) compact ($p_{l,moy}$ = 1 MPa)
	1.4 – 2.3		Alteration of vosgien clay	Compact clay (IP = 35), compact ($p_{l,moy}$ = 1 MPa)
		0.65–2.3	Molasse	Alternation of clayey marl (σ_c = 10 MPa) and sandstone (σ_c = 15 to 19 MPa)
Average	*7.4*	*6.9*		

Table 6.3. *Values of the dynamic unit friction in clayey ground*

It should be noted that several friction values for the same category of soil indicate the influence of other parameters such as the overcut, the height of the overburden, the trajectory corrections or the quantity of lubricant injected.

In the absence of lubrication, the unit friction f of sand and gravel reduces (7.4 kPa) to sand (5.4 kPa) and finally to clay (3.25 kPa). In the presence of lubrication, the nature of the soil does not seem to have as much importance.

In order to regroup all the unit frictions determined in this way within the general categories, the values are grouped together into 6 classes of soil whose general characteristics are defined below (see Table 6.4). The unit frictions f and f_{lub} within each class of soil are given in Table 6.5.

Soil class	γ (kN/m^3)	P_l (MPa)	c_u (kPa)	φ_u (°)	c' (kPa)	φ' (°)
1 – Coherent soft	17	0.3	20	0	10	17
2 – Granular loose	18	0.6	0	30	0	30
3 – Coherent quite stiff	19	1	40	5	20	20
4 – Granular quite compact	19	1.5	0	34	0	34
5 – Coherent stiff to hard	20	2.5	75	10	25	25
6 – Granular compact	20	3	0	38	0	38

Table 6.4. *Definition of the 6 classes of soil*

Class of soil	Not lubricated			Lubricated		
	f (kPa)	Number of linear	Values analyzed (ml)	f_{lub} (kPa)	Number of Linear	Values analyzed (ml)
1 – Coherent soft	–	–	–	–	–	–
2 – Granular loose	7.9	3	82	1.8	2	204
3 – Coherent quite stiff	5.6	2	44	3.1	2	153
4 – Granular quite compact	6.5	6	152	2.0	6	320
5 – Coherent stiff to hard	1.7	3	105	0.7	1	140
6 – Granular compact	3.1	3	110	6.9	4	210

Table 6.5. *Values of unit friction grouped together according to soil classes*

For granular soil, Table 6.5 shows that without lubrication, the more compact the soil is, the less significant the friction (from 7.9 kPa to 6.5 kPa then 3.1 kPa). This can be explained by the arching described by Terzaghi: the contact stresses exerted on the pipes reduce when the internal friction angle of the soil increases.

For cohesive soil, according to the compactness of the clay and thus the stability of the excavation, a friction without lubrication of the magnitude of 5 kPa for stiff clay can be obtained, which reduces to 2 kPa for stiff to hard clay.

When lubrication is done, the unit friction values, clearly lower, vary less according to the compactness of the soil. The higher values of class 6 can be essentially explained by an atypical site where the width of the overcut was abnormally low.

6.2.2.2. *Results of other studies*

Three statistical studies on the friction values involved during microtunneling in Japan, the United States and Norway can also be stated. These can be compared with the results obtained by the French National Research Project "Microtunnels" (see Table 6.6). These do not result from identical approaches and furthermore the use of lubricants is not specified:

– working group no. 3 of the Japan Society of Trenchless Technology (JSTT, 1994) has gathered 191 data elements for microtunneling with hydraulic mucking and 69 data elements with screw mucking. The unit friction here is an average conventional friction f_{conv} calculated from the final total thrust from which a conventional stress at the head has been subtracted. The values of the stress at the head are estimated in this approach by the first value of the jacking thrust;

– the Geological Laboratory of US Army Corps of Engineers has studied 12 microtunneling sections (Coller *et al.*, 1996). These studies did not take into account the stress at the head; the values obtained correspond to the average apparent friction f_{app}. The JSTT compared both approaches (f_{app} and f_{conv}), and concluded however that taking into account the total thrust, by neglecting the thrust at the head, leads to an overestimation of the average unit friction of 1 kPa ($f_{app} = f_{conv} + 1$);

– the Norwegian Geotechnical Institute has listed 40 microtunnel sites sunk in sand and sand and gravel (Lauritzen *et al.*, 1994). Like the previous study, this one does not take into account the impact of the thrust at the head and the corresponding values obtained correspond thus to the average apparent friction f_{app}.

Table 6.6 shows a summary of these results with the average, minimum and maximum values obtained during these studies for the three categories of soil considered previously.

The analysis of these results is not easy:

– firstly, due to the approximations made for the calculation of the unit friction: the Norwegian and American studies consider the average apparent friction f_{app}, whereas the Japanese studies take into account the average conventional friction f_{conv} which assumes a constant thrust at the head. However, the French National Research Project "Microtunnels" has shown that this varied significantly during the same section;

– secondly, the Japanese and American studies do not distinguish dynamic friction from static friction;

– finally, the lubrication conditions, which play a predominant role on the frictional forces , are not specified in these studies.

Ground	Unit friction (kPa)					
	PN Microtunnels		*JSTT* (f_{conv})		*G.L. of US Army*	*Norwegian Geotech. Inst*
	f	f_{lub}	Screw mucking	Hydraulic mucking	f_{app}	f_{app}
Clay	1.4 to 5.8 3.25	0.65 to 3.3 2.25	0.7 to 16 4.9	2.1	0.8 to 13 4.6	–
			3.5			
Sand	4.5 to 7 3 5.4	0.65 to 4.9 2.0	1.0 to 19 5.1	2.8	2.7 to 12 6.1	2.0 to 11 5.7
			4.0			
Sand and gravel	1.8 to 17 7.4	3 to 10 6.9	5 to 6.9 6.0	3.6	–	–
			4.8			

Table 6.6. *Summary of the minimum, maximum and average values of unit friction noted in the text (in kPa)*

However, by considering that taking into account the average apparent friction f_{app} leads to an overestimation of 1 kPa in the unit friction value as estimated by the JSTT, the values listed from the different studies seem very close. In the case of clay they will be 3.5 kPa and 3.6 kPa respectively for the Japanese and American studies. These values overall are situated between the values obtained in the French National Research Project "Microtunnels" with or without lubrication. All the results therefore seem coherent.

In addition, it is interesting to note that the JSTT has highlighted a unit friction twice as high for screw mucking as for hydraulic mucking without giving any explanation. It is possible that hydraulic mucking contributes to the lubrication and screw mucking leads to higher values of the thrust at the head.

6.2.3. *Calculation methodology for frictional forces*

We present here a method for calculating the frictional forces during soil-pipe contact. This method consisting of a bibliographical summary will then be compared to the experimental results.

The approach considered here examines in a first step the stability of the excavation made by the boring machine (see Figure 6.4):

– if this is unstable, the loosened ground comes into contact with the entire pipeline. The frictional forces F are then calculated by multiplying the contact stresses N exerted on the pipes by the friction coefficient μ characterizing the state of roughness of the soil-pipe interface;

– if the excavation is stable and if the consistency of the soil is less than the thickness of the annular space, the pipeline slides on its base, inside the open annular space. The frictional forces are equal to the product of the actual weight of the pipes and the soil-pipe friction coefficient;

– if the excavation is stable and if the consistency of the soil exceeds the width of the annular space, we return to the first case and calculate the frictional forces from the contact stresses exerted on the pipe.

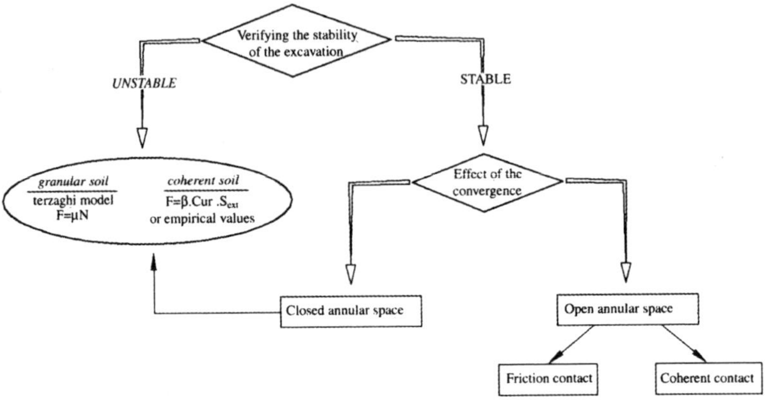

Figure 6.4. *Stability of the excavation*

6.2.3.1. *Verification of the stability of the excavation*

The English Pipe Jacking Association has suggested a method to estimate the confining pressure required (σ_T) to ensure the stability of the overcut.

a) In the case of cohesive soil, the short-term stability is linked to the undrained cohesion and the pressure that will be necessary to maintain a stable excavation is given by the following relation:

$$\sigma_T = \gamma.(H + \frac{D_e}{2}) - T_e.c_u$$

with:

– γ: specific weight of the soil above the pipes;

– H: height of the overburden over the pipes;

– D_e: diameter of the excavation;

– c_u: non-dewatered cohesion;

– T_c: stability coefficient of cohesive soil (values given in Figure 6.5 depending on H/De).

The excavation is stable if this internal pressure σ_T is less than or equal to 0.

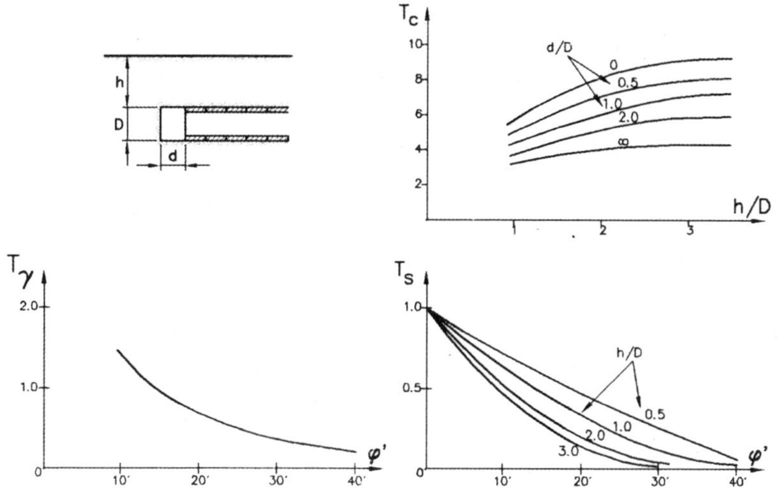

Figure 6.5. *Values of the stability coefficients* T_γ , T_c, T_s

b) In the case of non-cohesive soil, the stability depends on the internal angle of friction of the soil φ . In this case there is no simple general solution and the PJA proposes to consider the following two configurations for calculating σ_T:

– $\sigma_T = \gamma.D_e.T_\gamma$ in the absence of an overload above the pipeline, $T\gamma$ represents the stability coefficient indicated in Figure 6.5 depending on φ ;

– $\sigma_T = q_s.T_s$ in *the presence of a significant overload* q_s and at a low depth. The weight of the soil is then neglected and T_s represents the stability coefficient given in Figure 6.5 depending on φ and H/D_e.

Based on these relations when the soil is purely frictional, σ_T is always positive; as a result, the excavation is always unstable in the absence of a confinement pressure.

6.2.3.2. *Ground convergence effect*

The overcut represents the difference of radius between the excavation and the pipeline. Even in the case of a stable excavation, the ground can cave in on the pipes due to its flexible discharge. The vertical and horizontal decrease in the diameter of the excavation resulting in the flexible discharge of soil are calculated according to the state of initial stresses by adopting a law of flexible behavior of the ground, which leads to the following relations:

$$\Delta_V = \frac{1-v_s^2}{E_s}.D_e.(3.\sigma_V - \sigma_h)$$

and:

$$\Delta_h = \frac{1 - v_s^2}{E_s} . D_e . (3.\sigma_h - \sigma_v)$$

with:

- Δ_v reduction in the diameter of the excavation in the vertical direction,
- Δ_h reduction in the diameter of the excavation in the horizontal direction,
- V_s Poisson's coefficient of the soil,
- E_s Young's modulus of the soil.

If a pressure p is applied inside of the overcut (in a case where a lubricant is injected), this leads to a constant increase Δ_p in the excavation diameter:

$$\Delta_p = \frac{1 + V_s}{2.E_s} . p' . D_e$$

where p represents the effective internal pressure in the annular space (equal to the confinement pressure p reduced by the pore pressure of the ground).

According to the thickness of the overcut(s) made by the cutting wheel of the boring machine, two types of Figures can be shown:

– if Δ_v, and Δ_h – $\Delta_p < s$, there is not contact between the soil and the pipe, the annular space remains open and the friction is caused only by the boring machine's own weight;

– if Δ_v (and Δ_h) $\Delta_p \geq s$, there is contact between the soil and pipe, the annular space remains closed, and the frictional forces are linked to the stresses exerted by the soil on the pipes.

6.2.3.3. *Calculation of frictional forces for unstable excavation in granular soil*

The frictional forces are calculated by multiplying the total normal stress (N) that the soil exerts on the pipes by the frictional coefficient μ.

6.2.3.3.1. Determining the normal stress

The normal stress (N) acting on the outer surface of the boring machine is obtained by integrating the normal stress σ_n acting on a surface element dS. This is determined from the principal vertical (σ_v) and horizontal (σ_h) stresses of the ground. At a given point of the pipe, these stresses are given by (Figure 6.6):

$$\sigma_v = \sigma_{EV} + \gamma(D_e / 2 - y)$$
$$\sigma_h = K_2[\sigma_{EV} + \gamma(D_e / 2 - y)]$$

– σ_{EV}: vertical stress on the roof of the pipe,

– y: ordinate of the point P with respect to the centre of the pipe,

– K_2: thrust coefficient of soil acting on the pipe ($K_2 = 0.3$) (according to Stein, 1989).

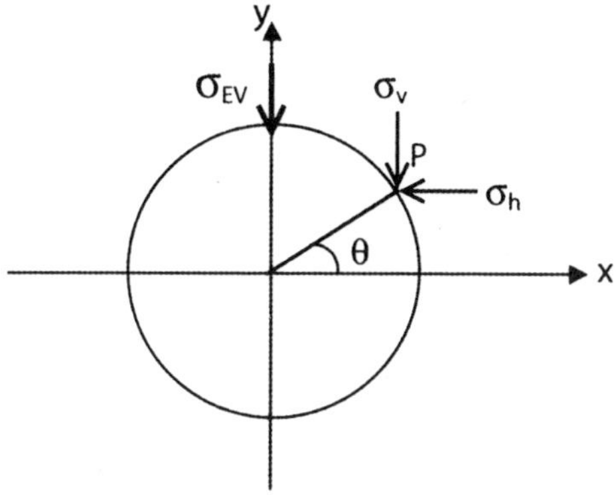

Figure 6.6. *State of stresses around the pipe*

The normal stress n acting on the surface of a pipe per linear meter, obtained by integration of the normal stress over the entire surface, is therefore defined as follows:

$$n = D_{ext}.\frac{\pi}{2}\left\{(\sigma_{EV} + \frac{\gamma.D_e}{2}) + K_2.(\sigma_{EV} + \frac{\gamma.D_e}{2})\right\}$$

where D_{ext} is the outer diameter of the pipe.

6.2.3.3.2. Determining σ_{EV}

Digging the microtunnel will disturb the initial state of the stresses around the excavation. This new state of stress, caused by the relaxation of the soil as a result of an overcut, can only be determined using a model. Studies carried until now as part of the National Project have shown that the Terzaghi model provides satisfactory results, close to actual values (Pellet, 1997, Phelipot, 2000).

The Terzaghi model (1951) assumes that the soil located above the pipe "slides" with respect to two vertical planes. These movements are sufficiently significant to lead to the creation of shear planes (see Figure 6.7).

The resolution of the differential equation of the equilibrium of a section of horizontal soil subject to shear stresses according to slip planes described earlier gives the expression of the vertical stress of the soil on the roof of the pipe:

$$\sigma_{EV} = \frac{b.(\gamma - \frac{2.c}{b})}{2.K.\tan\delta} \cdot (1 - e^{-2.K.\tan\delta.H/b})$$

with:

– H: Height of the overburden on the roof of the pipeline,
– γ : specific weight of the overlaying strata,
– K: horizontal pressure coefficient of the soil above the excavation,
– b: width of the affected ground,
– δ : friction angle of soil in place/decompressed soil above the excavation,
– c: cohesion of the soil.

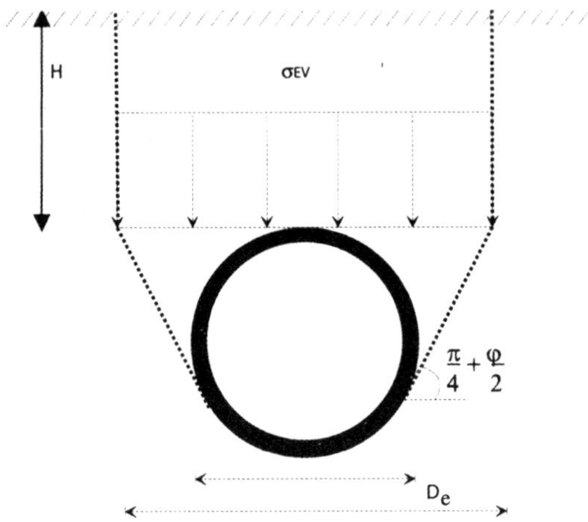

Figure 6.7. *Shear corners, Terzaghi model*

A certain number of assumptions (roughness of the slip planes, geometry of the shear corners) need to be considered in order to define the experimental parameters K, b and δ. Studies carried out to date (Pellet, 1997, FSTT, RS 25) have shown the adequacy of the model with the results obtained during follow-ups of the microtunneling sites, for parameters defined in the following manner:

$$K = 1 \quad \delta = \varphi \quad b = D_e[1 + 2\tan(\frac{\pi}{4} - \frac{\varphi}{2})]$$

This formula is slightly different from the basic Terzaghi model (see paragraph 6.4.3.1) because it has been derived from experimental results.

The vertical stress on the roof of the pipeline σ_{EV} can be represented by defining a coefficient k less than 1, which, by reducing the weight of the soil γ H, represents the arching of the ground (see Figure 6.8 and Figure 6.9):

$$\sigma_{EV} = k.\gamma.H$$

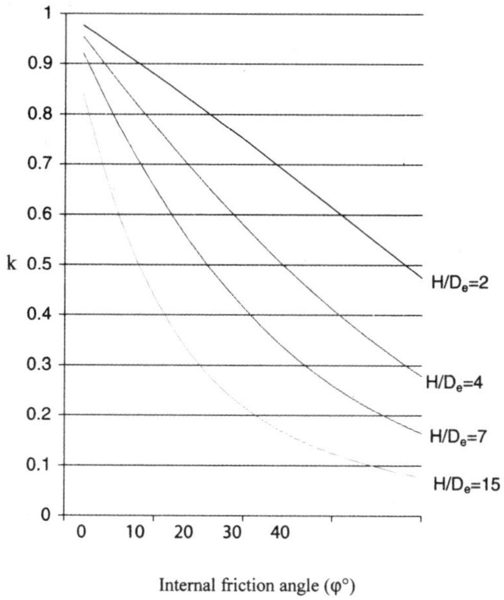

Figure 6.8. *Reduction coefficient k depending on φ*

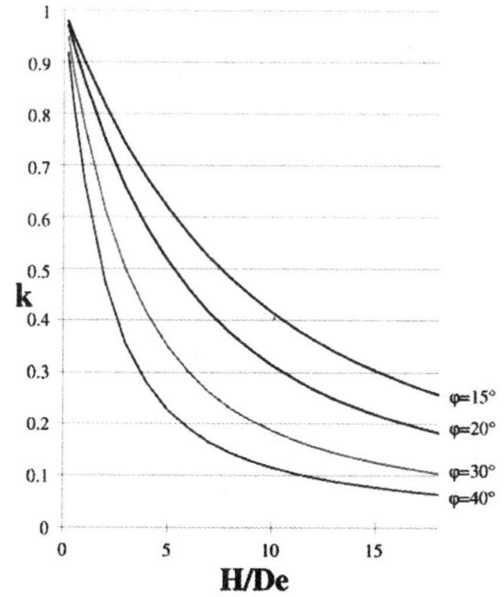

Figure 6.9. *Reduction coefficient k depending on H/De*

For granular soil, as the cohesion is zero, the coefficient k becomes:

$$k = \frac{1 - e^{-2.K.\tan\delta.H/b}}{2.K.\tan\delta.H/b}$$

Figure 6.8 and Figure 6.9 illustrate the variation of the coefficient k for granular soil depending on the ratio H/D_e and the internal friction angle of the ground φ.

It is accepted that when the height of the ground above the pipeline is low ($H/b < 1$), the decompression movements caused by the excavation act on the entire mass of the ground covering the microtunnel. The arching is thus neglected and the total mass of the soil above the pipeline is considered for the calculation of σ_{EV} (Szechy, 1970, AFTES, 1982).

6.2.3.3.3. Determining the frictional force

The frictional force is finally obtained by multiplying the normal stress N applied on the surface of the pipes by the soil-pipe frictional coefficient μ. The choice of the coefficient is discussed in paragraph 6.2.5.1.

6.2.3.4. *Calculation of frictional forces for unstable excavation in cohesive soil*

The shearing stresses caused by the contact between the clayey soil and the pipeline is dependent on the undrained cohesion c_u of the soil, of the coefficient β characterizing the soil-pipe adhesion (which depends on the type of pipe surface) and of the total surface of the pipes jacked. In fact, in the case of relatively large soil-pipe displacements caused by jacking, the clay in contact with the pipeline is greatly reworked and it is better to consider the undrained cohesion of reworked clay c_{ur}, that is:

$$F = \beta.c_{ur}.\pi.D_{ext}.L$$

The undrained cohesion in the reworked state c_{ur} can be estimated from the flow index $I_L = (w - w_p)/IP$ in the abacus presented in Figure 6.10 (Leroueil *et al.*, 1983). However, this approach assumes that the natural water content of clay in contact with the pipeline has not been modified by percolations of the mucking liquid or injection liquid.

The value of β, which characterizes the soil-pipe interface, was the object of various studies for piles (DTU 13.2). An average value of 0.6 will be considered (in the case of piles drilled in concrete of large diameter).

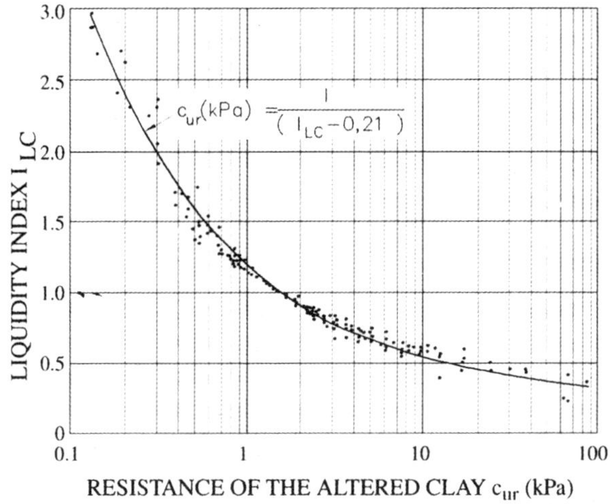

Figure 6.10. *Estimation of the undrained cohesion of reworked clay*

6.2.3.5. *Calculation of frictional forces for a stable excavation*

If the excavation remains stable but the convergence is such that the soil comes in contact with the pipeline, the conditions mentioned in paragraph 6.2.3.3 or 6.2.3.4 are encountered once more.

If the excavation remains stable and the convergence is less than the annular space, the pipe train slides on its base inside the open annular space. The frictional forces then depend on the nature of contact between the soil and the pipe.

6.2.3.5.1. For frictional soil

The frictional forces are equal to the product of the dead weight of the pipes and the soil-pipe frictional coefficient:

$$F = \mu.L.W$$

with:

- W: dead weight of the pipeline per linear meter,
- μ : soil-pipe frictional coefficient,
- L: Total length of the pipeline jacked.

In the case of a stable excavation located below the groundwater table (γ_w), or when the annular space is entirely filled with bentonite slurry (γ_b), the pipe string is subjected to buoyancy, which directly opposes its own weight. It is then advisable to consider the saturated weight of the pipeline. If it is negative, then the pipeline will float and the friction will act on the crown, hence the following general formula:

$$F = \mu.L.\left|W - \pi.\gamma_l.\frac{D_{ext}^2}{4}\right|$$

with:

- γ_l : equal to γ_w or γ_b depending on the case,
- D_{ext}: outer diameter of the pipes.

6.2.3.5.2. For cohesive soil

The instrumentation used at the actual site in England (Milligan, 1995) has shown that there exists a frictional type relation between the shear stress and the total normal stress in all types of soil except soft clay. Consequently, the formula related to frictional soil is used to calculate the frictional forces. This frictional behavior of most of the clayey soil is explained by Milligan and Norris (1999) with the help of an excavated surface, which is not smooth but relatively rough with sporadic and irregular contacts.

In highly special cases of soft clay, the theory of Haslem can be referred to, where the frictional forces result from the undrained cohesive soil (Haslem, 1986).

6.2.4. *Comparison of various approaches with experimental values*

The comparison of experimental results with results of calculation models helped establish a certain number of guidelines for the choice of calculation hypotheses.

6.2.4.1. *Calculations-measurements comparison: granular soil without lubrication*

The comparison of experimental results with calculation methods stated in paragraph 6.2.3 has shown us that in the absence of lubrication, the excavation done by the boring machine in sand or sand and gravel was not stable and that the soil-pipe contact stresses could be calculated from the relation proposed in paragraph 6.2.3.3. The frictional forces are obtained by supposing a soil-pipe frictional coefficient μ equal to 0.3, in accordance with the values given by Stein (Stein, 1989).

Only concrete pipes were the subject of experimental follow-ups in granular soil. It is possible that the pipes in vitrified clay or GRP may have a lower frictional coefficient.

Figure 6.11 compares the values resulting from the calculation with those deduced from experimental follow-ups. We note that the calculations lead in most cases to a slight overestimation of the unit friction. Only one point is clearly lower than the calculated value: this is a section driven in coarse alluvia; it is therefore possible that the soil-pipe frictional coefficient may actually be greater than the average value of 0.3 considered here for all the granular ground.

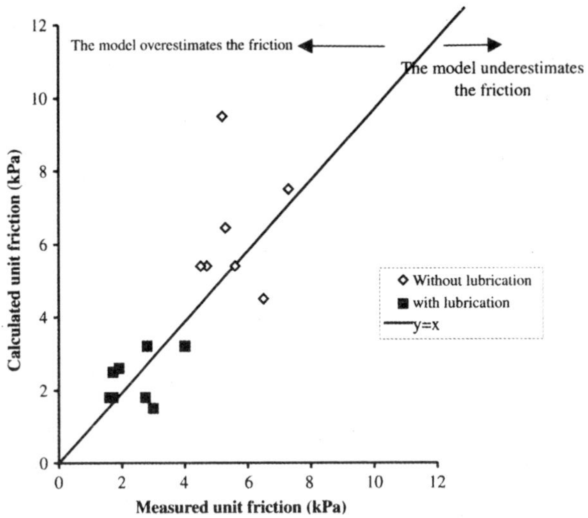

Figure 6.11. *Comparison of the calculated friction with the one obtained during experimental follow-ups: granular soil, case of an unstable excavation (μ = 0.1 for lubricated cases, and 0.3 for non-lubricated cases)*

6.2.4.2. *Calculations-measurements comparison: granular soil with lubrication*

When a lubricant is injected in the annular space, experimental results have shown that the friction values obtained were unpredictable, because of the disparity in the lubrication conditions: injected volume, continuous or discontinuous injection, etc. (see paragraph 3.1.3). It was also observed that injection of the lubricant could stabilize the excavation.

6.2.4.2.1. Non-stable excavation

For most sections jacked in granular soil, the injection of bentonite in the annular space did not help ensure the stability of the overcut. The friction can then be calculated by using the model based on the Terzaghi method and by applying a frictional coefficient taking into account the low frictional nature of the sand-bentonite mixture with respect to sand alone.

The results of the National Project have shown that when the soil closes in on the pipeline, the lubrication leads to a reduction in friction by about 50% to 77 % in granular soil, which cause the value of the frictional coefficient μ to vary between 0.07 and 0.15.

For a continuous injection and whose volume exceeds that of the annular space, the frictional forces may be calculated with a frictional coefficient of 0.1 (see Figure

6.11). The points of the Figure 6.11 show an underestimation of the friction with $\mu = 0.1$ corresponding to sites where the lubrication was not done continuously, with an injected volume less than the annular space.

6.2.4.2.2. Stable excavation

The results of experimental follow-ups have shown that on three sections or parts of sections, the injection of lubricating fluid helped ensure the confinement and the stability of the overcut.

At these sites, special attention was paid to lubrication. The volume of lubricant injected per meter jacked was particularly significant, of the order of 150 to 168 l/ml, i.e. 2 to 5 times the volume of the annular space, whereas the average of the injected volume was 70 l/ml over all the other sections (see paragraph 3.1.3.3).

The comparison between the deduced jacking values and those obtained by calculation described in paragraph 6.2.3.5 concerning a stable excavation and a frictional contact is presented in Figure 6.12. It appears that the model has the tendency to slightly underestimate the friction value.

It therefore seems necessary, in order to improve the estimation of the friction, to take into account the impact of trajectory deviations.

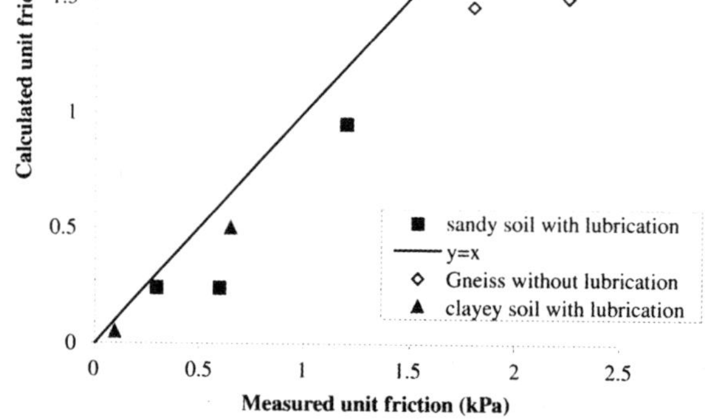

Figure 6.12. *Comparison of the calculated unit friction with the one obtained during experimental follow-ups: all types of soil, case of a stable excavation*

As the results of the National Project and the studies by Milligan and Norris (see paragraph 3.1.3.5) have shown, when the pipeline is jacked while being installed at its base, the horizontal trajectory deviations cause a non-negligible increase in frictional stresses. Based on the work of Milligan and Noris, this increase reaches 20 to 100% depending on the radius of curvature of the deviation and the length of the section.

The results of the National Project (see Figure 6.12) show that a multiplier equal to 1.5 applied to the results of the calculations leads to a correct estimation of observed values, except for a point corresponding to very low values of the friction.

6.2.4.3. *Calculations-measurements comparison: cohesive soil without lubrication*

Comparison of the results of the National Project with the calculation methods outlined in paragraph 6.2.3.1 showed that the excavation done by boring machines was stable in clayey ground. The calculations also showed that the convergence caused by the deconfinement led to the closing of the annular space. The soil-pipe contact stresses may then be calculated from the relation proposed in paragraph 6.2.3.4 by using the abacus of Figure 6.9 for determining c_{ur}.

The results of unit frictional forces deduced from experiments during the National Project are compared to the calculation in Table 6.7.

Site	BARR 3	BARR III	Champigny	Montmorency 3
Friction f measured (kPa)	1.4	1.35	6.3	5.8
Coefficient β	0.6	0.6	0.6	0.6
Cohesion c_{ur}	2	2	100	100
Friction f calculated (kPa)	1.2	1.2	60	60

Table 6.7. *Calculations-measurement comparison: cohesive soil without lubrication (closing of the annular space)*

It emerges from this comparison that the approach gives satisfactory results for the Barr site, situated in clay whose flow index is quite high. On the other hand, the values calculated are 10 times higher than the values measured for the Champigny and Montmorency 3 sites where the flow index is low, less than 0.1.

In these conditions, the value of the undrained cohesion in the reworked conditions is very high. On the other hand, in this clay that is quite stiff and of low water content, it is quite likely that the mucking water infiltrates into the annular space and considerably increases the water content of the reworked clay film in contact with the machine and then the pipes. This wetting may cause the C_{ur} value to drop considerably. It is also likely that the difficulties in mucking and filling of the stope have strongly limited this wetting process in the case of the sites at Barr.

6.2.4.4. *Calculations-measurements comparison: cohesive soil with lubrication*

At Geneva and on one portion of the section at Montmorency 3, the friction values are very small. It is likely that the injection of lubricant was sufficient to oppose the closing of the annular space. At Montmorency 3, following a significantly large injected volume (twice the volume of the annular space), the friction value momentarily reduced to 0.1 kPa. However, because of the discontinuity and irregularity of injections, the stability of the excavation could not be maintained over the entire section and the average friction was 3.3 kPa. For the Geneva section, sunk in stiff to hard cohesive soil (class 5), a lubricant volume equal to the volume of the annular space helped maintain the overcut.

For both these cases, the comparison of experimental results with those of the calculation method proposed in paragraph 6.2.3.5 (case of a stable excavation) shows quite a good compatibility (see Figure 6.12), the calculation under-estimating slightly the actual friction. It will therefore be suitable to similarly apply for granular soil a multiplier to help take into account the misalignments of pipes and therefore the "parasite" lateral friction.

Over the entire site at Montmorency 3 and Champigny, due to the irregularity of injections for one and because of the low volume injected for the other (less than the volume of the annular space), the ground closed in on the pipeline. In these conditions, the friction measured changed from 6.3 kPa to 2.8 kPa for Champigny and from 5.8 kPa to 3.3 kPa for Montmorency. This reduction may be attributed on the one hand to the lubricating effect which manifested itself by the reduction of the adherence coefficient β, and on the other hand to an additional humidification of the reworked clay layer in contact with the pipeline, leading to an additional reduction in c_{ur}.

6.2.5. *Guidelines for the calculation of pipe jacking stresses*

Following the summary of bibliographical elements and results of the French National Research Project "Microtunnels" concerning the observations at the site and the various calculation methods, we present below guidelines for estimating the pipe jacking stresses to be applied in the starting shaft of the boring machine.

These stresses result from dynamic frictional forces, which will vary with the jacked length, to which must be added the stresses relating to the additional friction caused by stoppages in jacking, as well as forces that are usually transmitted to the head of the boring machine to enable blasting and penetration.

6.2.5.1. *Dynamic friction: non-cohesive soil*

In the case of non-cohesive soil, the excavation is not stable and the ground always closes in on the pipeline (see paragraph 6.2.3.1). The calculation of frictional forces then consists of three stages:

Stage 1: determining the vertical stress σ_{EV} *at the roof of the pipeline*

This stress is calculated based on Terzaghi's silo theory, which leads to the following relation:

$$\sigma_{EV} = \frac{b(\gamma - \frac{2.c}{b})}{2.K.\tan\delta} \times (1 - e^{-2.K.\tan\delta.H/b})$$

with:

- H: height of the overburden at the roof of the pipeline,
- γ : specific weight of the overburden,
- K: earth pressure coefficient at rest,
- b: width of the affected ground,
- δ : friction angle of soil *in situ*/decompressed soil above the excavation,
- c: cohesion.

In general, it is recommended that we take:

$K = 1, \delta = \varphi$ and $b = D_e[1 + 2\tan(\pi/4 - \varphi/2)]$

Stage 2: determining the stress normal to the pipeline

The local normal stress n on one meter of pipe, obtained by integration of the normal stress over the entire surface from the σ_{EV} value calculated above, is given by the following relation:

$$n = D_{ext}.\frac{\pi}{2}\left\{(\sigma_{EV} + \frac{\gamma.D_e}{2}) + K_2.(\sigma_{EV} + \frac{\gamma.D_e}{2})\right\}$$

with:

- K_2: thrust coefficient of soil applied on the pipeline ($K_2 = 0.3$),
- D_{ext}: outer diameter of the pipeline,
- D_e: excavation diameter.

Stage 3: determining the frictional force

The local frictional force on the pipeline is obtained by multiplying the local normal force n by the value of the soil-pipeline friction μ. The total frictional force results from the integration of local stresses on the jacked length L, i.e.:

$$F = \int_0^L n.\mu.dl$$

For calculations we recommend the following values of the soil-pipe frictional coefficient μ:

– without lubrication: $\mu = 0.3$ in general and $\mu = 0.4$ for very coarse soil;

– with lubrication: $\mu = 0.1$ (continuous injection, volume injected > annular space);

– with lubrication: $\mu = 0.15$ to 0.2 (intermittent injection, volume injected < annular space).

NOTE: large injections (3 to 5 times the volume of the annular space) of special lubricating fluids well adapted to the ground (mixture of bentonite, polymers and microbeads, for example) may lead to very low values of μ or even the stabilization of the excavation, appreciably reducing the total frictional force. The calculation of frictional forces is then done according to the method proposed in section 6.2.5.2.2.

6.2.5.2. *Dynamic friction: cohesive soil*

In the case of cohesive soil, it is necessary first to check if the excavation is stable. The stability is ensured if the stress σ_T given by the relation below is negative or zero:

$$\sigma_T = \gamma.(H + \frac{D_e}{2}) - T_c.c_u$$

with:

– H: height of the overburden above the pipeline,

– D_e: excavation diameter,

– c_u: undrained cohesion,

– T_c: stability coefficient of cohesive soil (values given in Figure 6.5 depending on H/De).

If the excavation is unstable, the ground will close on the pipeline and the stresses will be calculated according to the procedure in paragraph 6.2.5.2.1.

If the excavation is stable, its elastic convergence is given by the following relations:

$$\Delta_v = \frac{1 - v_s^2}{E_s} D_e(3\sigma_v - \sigma_h)$$

and

$$\Delta_h = \frac{1-\nu_s^2}{E_s} D_e (3\sigma_h - \sigma_v)$$

with:

– Δ_v : reduction in the excavation diameter in the vertical direction (vertical convergence),

– Δ_h : reduction of the excavation diameter in the horizontal direction (horizontal convergence),

– ν_s: Poisson coefficient of the soil,

– E_s: Young's modulus of the soil.

There are two cases:

– if the convergence is greater than the annular space, the ground will close in on the pipeline and the forces will be calculated according to the method in section 6.2.5.2.1;

– if the convergence is less than the annular space, the pipeline will be installed in the excavation and the forces will be calculated according to the method in section 6.2.5.2.2.

6.2.5.2.1. Cohesive soil – excavation closing in on the pipeline

For an excavation closing in on the pipeline, the frictional force is given by the relation:

$$F = \pi . D_e \int_0^L \beta . c_{ur} . dl$$

The undrained cohesion in the reworked condition c_{ur} may be estimated from the value of the flow index I_L according to the abacus presented in Figure 6.10 (Leroueil *et al.*, 1983).

The value of β that characterizes the soil-pipe interface has been the subject of several studies for piles. For microtunneling, we can consider a coefficient β equal to 0.6 if the pipes are in concrete or 0.5 if they are in metal.

However, the comparison between calculated and measured values shows that the frictional forces between the pipeline and clayey ground are very sensitive to variations in the consistency of the reworked clay film in contact with the pipeline, in relation with the content of water either from mucking or from lubrication. In these conditions, particularly for clay with high consistency, it seems reasonable to base ourselves also on the empirical results presented in paragraph 6.2.2.

6.2.5.2.2. Cohesive soil – excavation not closing on the pipeline

In accordance with the results of Milligan (1995), we recommend adopting a friction type relation between the pipeline and the ground for all types of soil, except for soft clay (Haslem, 1986).

The frictional forces then result from the product of the pipes' own weight by the soil-pipe frictional coefficient:

$$F = \mu . L.W$$

with:

– W: own weight of the pipeline,

– μ : ground-pipe frictional coefficient,

– L: total length of the pipeline jacked.

In the case of a stable excavation located under the unconfined groundwater, or when the annular space is completely filled with bentonitic clay, the pipe string is subjected to buoyancy, which directly opposes its weight. It is then suitable to take into account the weight out of water of the pipeline. If this becomes negative, the pipeline will float and the friction will take place along the crown, hence the following general formula:

$$F = 1.5\mu . L . \left| W - \pi . \gamma_w \, D_{ext}^2 / 4 \right|$$

γ_w being the specific weight of the water or the slurry if the pipe is entirely lubricated.

The values of the frictional coefficient μ to be taken into account are:

– without lubrication: $\mu = 0.2$,

– with lubrication: $\mu = 0.1$.

The coefficient of 1.5 helps take into account the horizontal trajectory deviations. These, in fact, cause additional friction on the lateral walls of the excavation and not only on the base of the excavation (see paragraph 6.2.4.2).

6.2.5.3. *Additional friction caused by stoppage in jacking*

The additional friction is a function of the ground and particularly its propensity for creeping. It is also linked to the duration of the stoppage in jacking and we have had boring machines blocked after an extended stoppage of a weekend! The results obtained during the National Project have shown that this additional friction is noticeably proportional to the logarithm of the duration of the stoppage (see section 3.1.3.3).

We provide as an example the values recorded at a site built in marl with low sand content:

	Stoppage of a weekend	Stoppage of one night	Stoppage < 3 hours
f_{sup} (kPa)	2.4 kPa	1 to 2 kPa (*)	0.6 to 0.8 kPa

(*) It must be noted that the injection of a sufficient volume of lubrication fluid in the annular space helps oppose the tightening of the ground on the pipeline and reduce the amplitude of the additional friction. Thus, after an interruption of one night and depending on the lubrication conditions, the additional friction may vary between 1 and 2 kPa.

For a microtunnel of 800 mm outer diameter, and for a jacked length of 100 m, a stoppage of a weekend would correspond to an additional thrust of 600 kN, at times sufficient to cause blocking of the pipeline.

6.2.5.4. *Stress on the cutter head*

The stress on the cutter head depends essentially on the nature of the ground, the drilling diameter and the drilling parameters (jacking speed, mucking flows).

The stress on the head R_p may be characterized by r_p, apparent resistance at the head (see paragraph 3.1.4), which includes the effect of stresses on the cutting wheel and that of the mucking liquid pressure, i.e.: $R_p = r_p \cdot \pi \cdot \frac{D_e^2}{4}$

The experimental results have given maximum and average values of r_p that we have classified according to three groups of soil: sand, sand and gravel, and clay.

	Sand	Sand + Gravel	Clay
Max of r_p max	1800 kPa	2300 kPa	800 kPa
Average of r_p max	1000 kPa	1700 kPa	600 kPa

If we assume that the maximum values correspond to inadequate jacking situations, we will adopt average values for sizing the jacking system and the pipes, by noting that the stress corresponding to the active earth pressure represents a lower limit in relation to the risk of over-excavation at the face.

6.2.5.5. *Estimate of the maximum pipe jacking stress*

Experimental follow-ups of the National Project have shown that for sections of small length, the thrust peaks were caused by an increase in the thrust at the head. The maximum total thrust may then be estimated by the following relation:

For sections < 70 m $P_{tot} = F + R_p = r_p.\pi.D_e^2/4 + \int f.dS$

Beyond this distance, the largest thrust peaks are generally caused by the increase in friction during restarting. We can thus refine the estimate of the total thrust by comparing the previous relation with sup $\int(f + f_{sup}).ds$ and by retaining the greater of the two values, i.e.:

$$\text{For sections} > 70 \text{ m } P_{tot} = Sup\left\{r_p.\pi.D_e^2/4 + \int f.dS \quad ; \quad \int(f + f_{sup}).dS\right\}$$

6.3. Calculation of the maximum acceptable thrust by the pipes during jacking

6.3.1. *Calculation principle*

When the pipes are installed by jacking, they are subjected to transverse actions (weight of earth, overloads, etc.) and to longitudinal actions (thrust stresses). It is the latter that most often determine the sizing of pipes.

The maximum permissible thrust depends on the compressive strength of the material constituting the pipes, the contact surface between the pipes and also the straight angulations over the assemblies that cause an off-centering of thrust stresses. In fact, during jacking, the pipes, constituting an articulated train, are not aligned with one another, particularly because of trajectory corrections.

The distribution rings being able to transfer only compressive forces, the maximum thrust of jacking cylinders must take into account the off-centering of the application of the thrust.

The pipe manufacturers apply a safety coefficient that reduces the compressive strength of the material and thus defines the permissible stress on the pipes, which must not be exceeded on the site. This permissible thrust is generally defined up to a limiting angulation of 1°.

The German standard ATVA 161, adopted by the European and French standard NF EN 295-7 (January 1996) for clay pipes, proposes the application of a safety coefficient equal to the ratio of stresses σ_{max} / σ_0 (σ_{max} is the maximum marginal stress and σ_0 the stress distributed uniformly in the case of a resulting centred theoretical thrust).

The ratio of stresses σ_{max} / σ_0 is determined by the eccentricity of the thrust and, as long as its resultant acts outside the centre of the transverse section, it is characterized by the opening of joints on one side of the pipe. The importance of the opening of joints is provided by the ratio Z/d_a. The relation between Z/d_a and σ_{max} / σ_0 is presented in Figure 6.13, where Z represents that part of the diameter where there is compression in the contact plane.

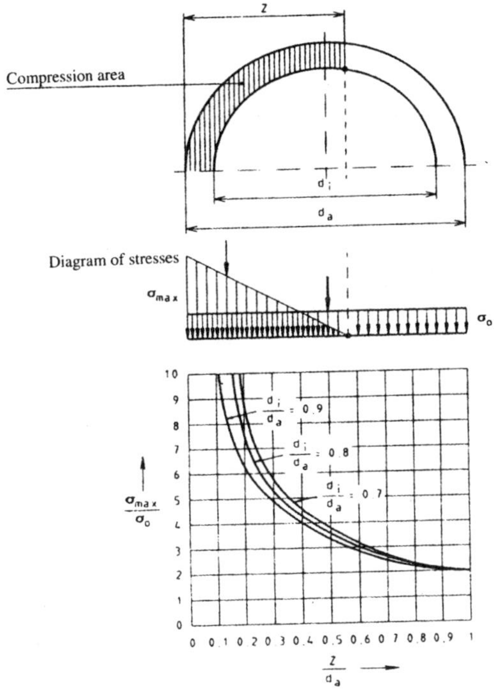

Figure 6.13. *Relation between the ratio* σ_{max} / σ_0 *and the eccentricity ratio* Z/d_a *(part of the joint section transmitting the thrust), according to the ATVA standard (Stein* et al., *1989) and NF EN 295-7 (January 1996)*

When we consider that the axial stress is still transmitted by the entire surface of the joint ($Z/d_a = 1$), which corresponds, according to the German standard, to a limiting angulation of 0.5°, the σ_{max} / σ_0 ratio is equal to 2.

The standard NF EN 295-7 suggests taking into account this value in the common cases, that is to say when the trajectory corrections occurring on inclined paths and linear directions are involved.

For larger deviations, produced by opening of joints in the case of projects with curved paths, $Z/d_a < 1$, the importance of the eccentricity in individual joints is taken into account by using the minimum value of Z when determining σ_{max} / σ_0 of Figure 6.13.

The maximum permissible thrust on the pipes is calculated using the following relation:

$$P_{max} = \frac{\sigma_0}{\sigma_{max}} \times \frac{P_{theoretical}}{f}$$

The theoretical thrust, declared by the manufacturer depends on the resistance to jacking ($R_{jacking}$ = Axial compressive strength of the material/Annular surface of the joint) and the characteristics of the waterproofing element and the distribution ring. The safety coefficient f is fixed at 1.6 according to NF EN 295-7 for automatic control jacking machines and at 2 for manually driven systems.

The Pipe Jacking Association also determines the permissible thrust according to the portion of the joint that transmits the stress. This is calculated based on a linear behavior model of the joint. The permissible jacking thrust is linked to the thickness of the joint; it is given by Figure 6.14 up to a maximum angulation of 1°. We observe that for an angulation of 0.5°, the "equivalent safety coefficient" varies between 1.6 and 2; and it reaches 2.5 to 3 for an angulation of 1°.

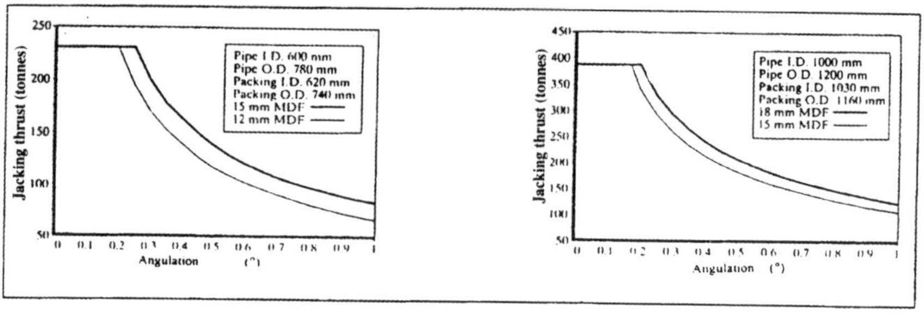

Figure 6.14. *Permissible thrust on the ends of the pipes according to the angulation at the level of the joint, according to the Pipe Jacking Association, 1995*

As far as the bending loads are concerned, the amplitude of angulations between pipes during jacking and the importance of traction generated in the pipes are not yet well known. It is therefore important to be able to control trajectory deviations during jacking in order to limit the angulations between the pipes and as a result the concentration of stresses at the joints.

6.3.2. *Permissible stress in the pipes*

The verification of axial stresses consists in comparing the maximum stress in the pipe (σ_{max}) with the permissible stress, which varies according to the materials used (see Table 6.8).

The permissible stress is the value which must not be exceeded on the site. This limitation is defined in order to take into account uncertainties regarding materials, the geometry of pipes and the difficulty in estimating the angular deviations between pipes and the off centering of the axial thrust.

	Tensile strength (MPa)	Permissible stress (MPa)
Concrete	25 to 45	10 to 18
High performance concrete	45 to 80	18 to 32
GRP	90	30
Sandstone	30 to 35	15 to 18
Steel	360	240

Table 6.8. *Permissible compressive stresses (guide values) in the pipes*

NOTE: the values indicated for pipes in GRP and clay pipes were obtained by manufacturer tests. The values indicated for concrete pipes are defined by tests and are prescriptive.

6.4. Calculation of the cross-section of pipes

6.4.1. *Various verifications of the calculation of the size of pipes*

There exist essentially two sizing situations for pipes:

– *during the work*, in the course of jacking operations, where the pipes are essentially subjected to axial loads caused by jacking stresses, possibly off-centered because of possible misalignments;

– *during operation*, under the effect of various transverse loads being exerted on the buried pipes: ground loads, external or internal water loads, surface overloads (permanent or traffic).

The first case, where experience has shown that it is generally the situation that sizes the pipes, was previously mentioned. It consists in estimating the maximum pushing stresses that may be exerted depending on the length of section, the ground-pipe friction and the stress at the head (see section 6.2), and in taking into account the off-centering effects of these stresses following deviations (see section 6.4). We will not reconsider these matters here.

We will limit ourselves here to examining the second case for sizing of pipes under transverse loads. There currently exists no French regulation for the sizing of pipes installed "without trenches"; the fascicle 70 of the General Technical Specifications deals in fact only with the case of structures constructed traditionally in trenches. It will however serve as a guide for the following guidelines, and which will be adapted to the specifications for trenchless work.

6.4.2. *General calculation principles: basic Terzaghi model*

The digging of a cavity in the soil, such as a microtunnel, disturbs the initial condition of the stresses in the ground situated around the pipeline, because of the relaxation of the ground due to the opening of the cavity. The new state of the stresses may be determined using a behavior model.

The model presented hereinafter, due to Terzaghi, supposes that the ground located above the pipeline slips in relation to two vertical planes separated by a width b, depending on the pipeline diameter and the mode of "rupture" envisaged, and which will subsequently be specified (see Figure 6.15). An arching is then created.

Taking into account the shear stresses along these planes, we can formulate the equilibrium of a section of width b along the vertical direction:

$$\gamma.b.dz + \sigma_V.b - (\sigma_V + d\sigma_V).b - 2.(c + \sigma_V.K.\tan\delta).dz = 0$$

with:

- γ : specific weight of the ground (kN/m^3),
- c: cohesion (kPa),
- δ : frictional angle on the vertical planes,
- $\tau = c + \sigma_v.K.\tan\delta$ [kPa] (Mohr-Coulomb criterion),
- K: horizontal pressure coefficient of the soil, equal to the ratio of horizontal and vertical stresses.

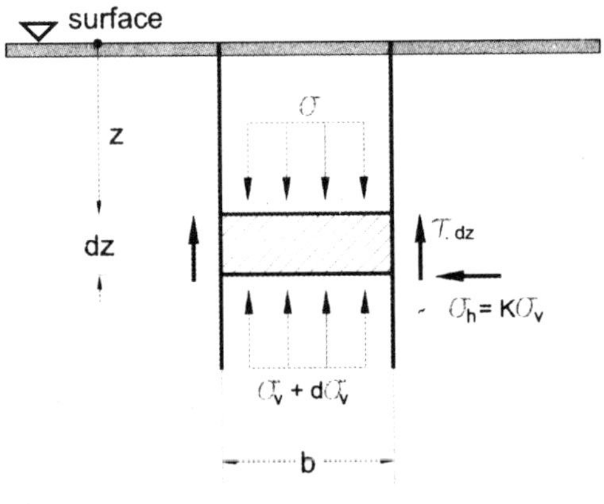

Figure 6.15. *Equilibrium of a horizontal section of the ground – Terzaghi Model*

We obtain the following differential equation:

$$\frac{d\sigma_v}{dz} = (\gamma - \frac{2.c}{b}) - \frac{2.K.\sigma_v.\tan\delta}{b}$$

The solution of this equation for z = H (crown of the pipeline) may be written as:

$$\sigma_{EV} = \frac{H.(\gamma - \frac{2.c}{b})}{2.K.\tan\delta.\frac{H}{b}}(1 - e^{-2.K.\tan\delta.H/b})$$

Using the notation $\sigma_{EV} = k.\gamma.H$, we introduce the coefficient k, which helps take into account the arching of the ground. If the overlying ground is not cohesive, the coefficient k becomes:

$$k = \frac{1 - e^{-2.K.\tan\delta.H/b}}{2.K.\tan\delta.H/b}$$

The parameters H and γ may be easily determined. On the other hand K, b and δ are parameters that must take into account a certain number of additional considerations.

6.4.3. *Vertical loads to the soil alone*

6.4.3.1. *The experimental Terzaghi model*

In the Terzaghi model, the width "b" is calculated taking into account the formation of two shear corners sloping at $\frac{\pi}{4} + \frac{\varphi}{2}$ with respect to the horizontal (see Figure 6.16), i.e.:

$$b = D_e.\tan\left(\frac{3\pi}{8} - \frac{\varphi}{4}\right)$$

D_e being the outer diameter of the pipeline, the friction angle TM is taken as being equal to ϕ the internal friction angle of the ground.

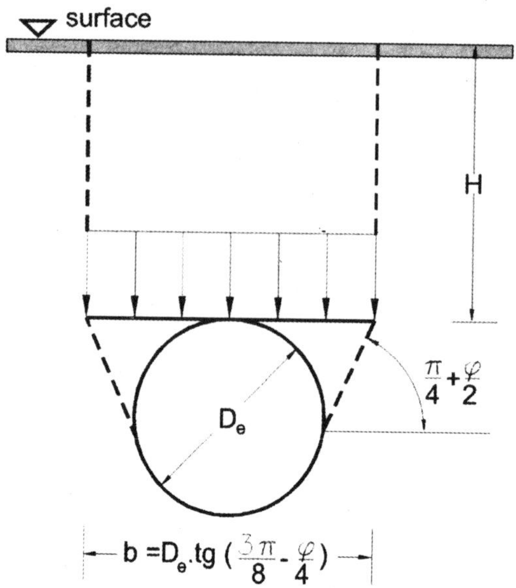

Figure 6.16. *Formation of shear corners – Terzaghi Model*

The physical analysis of the phenomenon leads to the two following observations:

– when the overburden of the ground above the pipeline is small (H ≤ b), the decompression movements caused by jacking risk affecting the total mass of the ground covering the pipeline and the arching is generally neglected;

– when the pipeline is driven at a large depth (H > 2.5 b), the arching will not extend beyond a distance of 2.5 b above the pipeline key.

We therefore express the vertical stress of the ground as follows:

– if $H \leq b$:

$$\sigma_{EV} = \gamma.H$$

– if $b < H < 2.5.b$:

$$\sigma_{EV} = \frac{b.(\gamma - \frac{2.c}{b})}{2.K.\tan\varphi}(1 - e^{-\frac{2.K.\tan\varphi.H}{b}})$$

- if $H > 2.5.b$:

$$\sigma_{EV} = \frac{\gamma.b - 2c}{2.K.\tan\varphi}$$

The value K was determined experimentally for sand: $K \approx 1$.

It must be noted that this method is applied in Japan with the values of b, ϕ and K proposed by Terzaghi.

6.4.3.2. *The ATV A161 method*

The calculation of pipes by microtunneling according to the German standard ATV A161 is an adaptation of the Terzaghi model. The width b is determined by: $b = D_e\sqrt{3}$, which corresponds to an angle $\varphi = 30°$ (see Figure 6.17).

The ground-backfill frictional angle is taken as being equal to $\delta = \frac{\varphi}{2}$, φ being the internal frictional angle of the ground. The horizontal pressure coefficient of the ground on the slip planes is equal to K = 0.5.

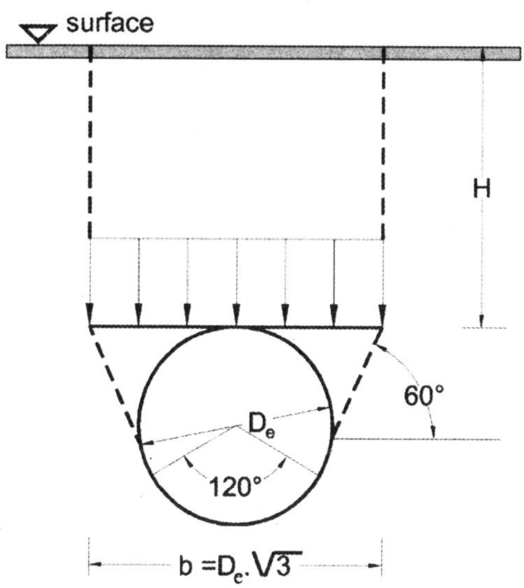

Figure 6.17. *Formation of shear corners. Model ATV A161*

6.4.3.3. *Leonards' model*

Leonards' formula is established for pipelines installed in trenches, based on the Marston theory for a trench equal to the outer diameter of the structure (see Figure 6.18). This theory is based on the same principles as that of Terzaghi, but the shear planes considered are vertical planes tangent to the generatrix.

The formulation is therefore the same as that of Terzaghi by considering $b = D_e$.

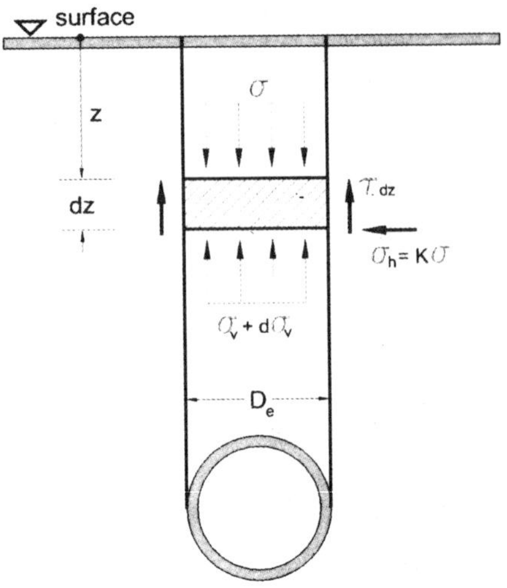

Figure 6.18. *Equilibrium of a horizontal section of the ground: Leonards' formula*

6.4.3.4. *Guidelines for the calculation of vertical loads*

We present below the guidelines for the sizing of pipes by basing ourselves on previous approaches and taking into account the influence of various physical parameters on the one hand, and the various conditions for implementation on the other.

We could use the Terzaghi or Leonards' method, both of which take into account the arching of the ground situated above the pipeline. However, it is recommended that the cohesion term be taken into account with great care:

– when the overburden H is low, the decompression movements caused by the construction of the structure risk affecting the entire enclosing ground above, especially when the ground is saturated with water; we will not take into account the cohesion in this case;

– for large overburden heights, the cohesion effect is durable and it is justified to take it into considering in the calculations. However, in the cases where $H > b$ and to be safe, we will make sure that the vertical pressure calculated with the cohesion is not less than $\gamma .D_e$.

It is advisable to be careful regarding the cohesion value c to be taken into account. Particularly in non-homogenous ground, we will take into account the material whose ϕ and c characteristics lead to the maximum value of the vertical pressure.

In the case of clay or highly plastic marl, measurements done on old structures show that the pressures increases slowly with time till they reach the initial geostatic pressure. This geostatic pressure is the pressure caused by the weight of the earth on top of the structure, when we do not take into account the lightening due to cohesion and frictional forces along the imaginary fracture plane. In these cases, in addition to other calculations and during the final phase, we will have to verify the resistance of the structure with a vertical pressure: $\sigma_{EV} = \gamma.H$ and a horizontal thrust $\sigma_h = K.\sigma_{EV}$. with K close to 1.

We recommend the application of the following method:

a) In the case of homogenous ground above the pipe, if the cohesion c and the internal friction angle are known with accuracy:

– calculate the vertical pressure σ_{EV1}

- if H > b :

$$\sigma_{EV1} = \gamma.b = \gamma.D_e.\tan(\frac{3\pi}{8} - \frac{\varphi}{4})$$

- if H < b :

$$\sigma_{EV1} = \gamma.H$$

– calculate the vertical pressure σ_{EV2} or σ_{EV3} given by the Leonards' formula, wherein kM is the Marston coefficient in a narrow section (width of the section equal to the diameter of the pipeline), and this in one or the other of the following cases:

- if c = 0:

$$\sigma_{EV2} = k_M.\gamma.H$$

- if c ≠ 0:

$$\sigma_{EV3} = k_M.H.(\gamma - \frac{2.c}{D_e})$$

with:

$$k_M = \frac{1 - e^{-2.K.\tan\varphi.\frac{H}{D_e}}}{2.K_a.\tan\varphi.\frac{H}{D_e}}$$

and:

$$K_a = \tan'^2(\frac{\pi}{4} - \frac{\varphi}{2})$$

– retain the vertical pressure σ_{EV} such as:

- if $\sigma_{EV1} > \sigma_{EV2}$:

$$\sigma_{EV} = \sigma_{EV2} \text{ (even if } c \neq 0)$$

- if $\sigma_{EV1} < \sigma_{EV3}$:

$$\sigma_{EV} = \sigma_{EV3}$$

- if $\sigma_{EV3} < \sigma_{EV1} < \sigma_{EV2}$:

$$\sigma_{EV} = \sigma_{EV1}$$

b) In the case of heterogeneous ground above the pipe or if the characteristics of the homogenous ground are not known well.

When the characteristics of the ground are not known precisely, we will take into account a basic value of 30° for the friction angle, the cohesion being taken as zero as a precaution. The vertical pressure is then given by the following relation, where kM is the Marston coefficient defined earlier:

$$\sigma_{EV3} = k_M.\gamma.H$$

c) If the host ground is made up of clay or very plastic marl.

In this case, and in addition to the previous calculations, we must undertake a verification of the pipeline with:

$$\sigma_{EV} = \gamma.H$$

6.4.4. *Horizontal loads of the ground*

The lateral pressure is effectively exerted on the structure only if the type of structure and the injections are such that the contact between the ground and the coating is ensured with sufficient effectiveness: the residual spaces are then very few and regularly distributed.

The lateral pressure is proportional to the vertical pressure and we assume it as being constant on the height of the structure. It is given by the following expression where q_0 is the effect of an overstress whose calculation principle is given in paragraph 6.5.5 (see Figure 6.19):

$$\sigma_h = K.(\sigma_{EV} + \frac{D_e}{2}\gamma + q_0)$$

where:

- σ_h : horizontal stress of the ground at the height of the centre of the pipeline,
- σ_{EV} : vertical stress of the ground at the level of the upper generatrix of the pipeline calculated previously.

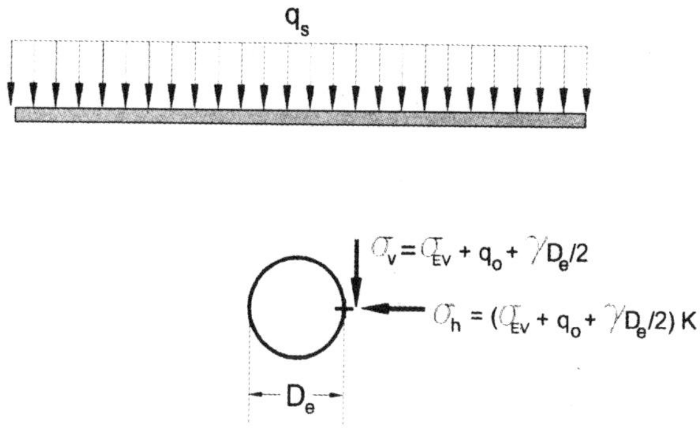

Figure 6.19. *Action of the active earth pressure*

The active earth pressure coefficient K may be given by:

– the Rankine formula: $K_a = \tan^2(\frac{\pi}{4} - \frac{\varphi}{2})$ active earth pressure coefficient;

– the AFTES recommendations (1976): $0.9 - \sin\phi < K < 1 - \sin\varphi$, this last value corresponding to the rest earth coefficient.

We therefore recommend, in the case of normally consolidated ground and, as a precautionary measure, the calculation of the earth pressure by the Ranking formula by limiting K to a maximum value of 0.5. However, for plastic clay or marl, we will consider K = 1, associated to the total vertical stress γH .

6.4.5. *Surface loads*

The application of surface loads will increase the vertical stress, as well as the horizontal stress on the right of the pipe. We distinguish between:

– permanent surface loads caused by the existence of backfills, buildings, etc.,

– intermittent surface loads caused by the traffic (road, rail or air).

6.4.5.1. *Permanent surface loads*

Permanent surface loads are generally caused by construction loads directly below or close to the pipeline. In the case of skewed loads, a special calculation will have to be done. An overstress q_s of large dimensions causes an increase in the vertical stress at the depth of the pipe, which can be calculated using the Terzaghi method and which is expressed as (see Figure 6.20):

$$q_o = q_s . e^{-2.K.\tan\delta.H/b}$$

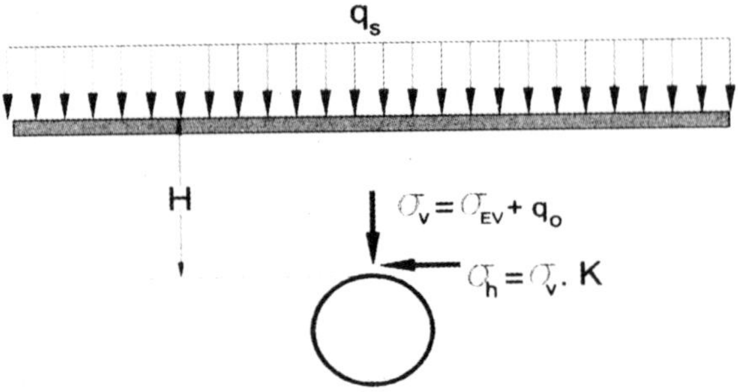

Figure 6.20. *Influence of a distributed overstress q_s*

This calculation method is the one adopted by Fascicle 70 that proposes:

$$K.\tan\delta = 0.15$$

For microtunneling, we recommend:

$$\delta = \phi \qquad b = D_e \tan\left(\frac{3\pi}{8} - \frac{\varphi}{4}\right) \qquad K = \tan^2\left(\frac{\pi}{4} - \frac{\varphi}{2}\right)$$

6.4.5.2. *Traffic loads*

Traffic loads may be caused by road, rail or exceptional traffic. These surface loads are statutorily defined.

6.4.5.2.1. Road traffic

Fascicle 70 defines the road loads that need to be taken into account for the calculation of sanitation work. They correspond to the most unfavorable type Bc loading system of the train defined by Fascicle 72-21 bis of the CPC, modified by impact factors. The overstress value q_o applied on the pipe is given by Table 6.9.

6.4.5.2.2. Rail traffic

The SNCF (French National Railway Company) defines rail loads to be taken into account for the calculation of buried pipelines in Appendix 1 of Fascicle NG EF 9C5 no. 1. Figure 6.21 gives the value of the overstress q_o depending on the depth of the pipeline and the existence of one or more tracks.

The impact coefficient to be used to account for the effects of the impact factor is:

$M_d = 1 + \frac{0.6}{H}$, where H is the depth of the pipeline. The value of the overstress $q_{o,d}$ to be taken into account on the pipeline is: $q_{o,d} = M_d \cdot q_o$.

Depth in m	Nominal Diameter (mm)												
	150	200	250	300	400	500	600	700	800	900	1,000	1,100	1,200
0.08	72.58	72.29	71.93	71.49	70.42	69.11	67.62	65.98	64.22	62.37	60.47	58.54	56.61
1.00	57.55	57.39	57.19	56.94	56.33	55.57	54.67	53.66	52.57	51.39	50.17	48.90	47.31
1.20	47.30	47.20	47.08	46.93	46.54	46.09	45.51	44.87	44.16	41.40	42.58	41.74	40.87
1.40	40.03	39.98	39.89	39.79	39.54	39.24	38.86	38.45	37.98	37.46	36.92	36.34	35.75
1.60	34.68	34.63	34.58	34.51	34.35	34.14	33.90	31.61	33.29	32.94	32.56	32.16	31.74
1.80	30.58	30.55	30.51	30.47	30.35	30.22	30.04	29.84	29.62	29.37	29.10	28.82	28.52
2.00	27.34	27.32	27.30	27.26	27.18	27.08	26.96	26.82	26.66	26.48	26.29	26.08	25.86
2.00	24.71	24.69	24.67	24.65	24.59	24.52	24.42	24.32	24.20	24.07	23.93	23.78	23.62
2.40	22.51	22.50	22.48	22.46	22.42	22.37	22.30	22.22	22.13	22.03	21.92	21.81	21.68
2.60	20.63	20.63	20.62	20.61	20.57	20.53	20.48	20.42	20.34	20.27	20.18	20.10	20.00
2.80	19.02	19.01	19.01	18.99	18.97	18.94	18.90	18.85	18.79	18.74	18.66	18.59	18.51
3.00	17.61	17.59	17.59	17.50	17.57	17.54	17.50	17.46	17.42	17.37	17.32	17.26	17.19
3.50	14.72	14.71	14.70	14.70	14.69	14.67	14.65	14.63	14.60	14.57	14.54	14.50	14.46
4.00	12.50	12.50	12.49	12.49	12.48	12.46	12.46	12.44	12.42	12.39	12.38	12.32	12.37
4.50	10.74	10.74	10.74	10.74	10.74	10.72	10.71	10.70	10.68	10.67	10.66	10.63	10.61
5.00	9.32	9.34	9.33	9.32	9.32	9.30	9.30	9.30	9.28	9.27	9.26	9.24	9.23
5.50	8.15	8.18	8.16	8.17	8.16	8.16	8.14	8.14	8.13	8.12	8.12	8.10	8.09
6.00	7.20	7.20	7.19	7.20	7.19	7.18	7.19	7.18	7.18	7.17	7.17	7.15	7.14

Depth in m	Nominal Diameter (mm)											
	1,300	1,400	1,500	1,600	1,800	2,000	2,200	2,500	2,800	3,000	3,200	3,500
0.80	54.68	52.78	50.94	49.14	45.78	42.79	40.21	37.13	34.99	34.06	33.44	32.92
1.00	46.31	45.02	43.74	42.49	40.12	37.96	36.06	33.71	32.00	31.17	30.55	29.90
1.20	39.99	39.10	38.22	37.34	35.68	34.09	32.67	30.88	29.48	28.76	28.17	27.46
1.40	35.14	34.52	33.90	33.28	32.06	30.92	29.86	28.46	27.31	26.68	26.14	25.44
1.60	31.31	30.88	30.43	29.98	29.10	28.24	27.43	26.34	25.38	24.83	24.34	23.68
1.80	28.22	27.90	27.58	27.24	26.58	25.94	25.31	24.43	23.64	23.16	22.72	22.10
2.00	25.63	25.40	25.16	24.91	24.42	23.91	23.42	22.70	22.04	21.62	21.23	20.67
2.20	23.44	23.26	23.08	22.90	22.50	22.11	21.72	21.14	20.58	20.22	19.86	19.35
2.40	21.55	21.42	21.27	21.13	20.82	20.50	20.18	19.70	19.22	18.90	18.60	18.14
2.60	19.90	19.78	19.67	19.56	19.31	19.06	18.79	18.38	17.97	17.70	17.42	17.02
2.80	18.43	18.34	18.26	18.16	17.96	17.74	17.52	17.18	16.82	16.58	16.34	15.98
3.00	17.13	17.06	16.98	16.90	16.74	16.56	16.37	16.07	15.77	15.55	15.34	15.02
3.50	14.42	14.37	14.32	14.26	14.15	14.03	19.90	13.70	13.47	13.32	13.16	12.91
4.00	12.29	12.26	12.22	12.18	12.10	12.02	11.93	11.78	11.61	11.49	11.37	11.18
4.50	10.59	10.57	10.54	10.51	10.46	10.39	10.32	10.21	10.08	9.99	9.90	9.75
5.00	9.22	9.19	9.18	9.15	9.11	9.06	9.01	8.92	8.82	8.75	8.68	8.57
5.50	8.08	8.06	8.05	8.03	8.00	7.96	7.92	7.85	7.78	7.72	7.66	7.57
6.00	7.14	7.13	7.11	7.10	7.07	7.04	7.01	6.95	6.90	6.85	6.81	6.74

This table is prepared for wall thickness that are: D_e(outer) = 1.2 D(inner)

Table 6.9. *Overstress q_o at the pipe due to moving loads in kPa (dynamic coefficients included)*

D: Nominal diameter of the pipe or range of the flexible metallic nozzle
B: Nominal width of the ovoid

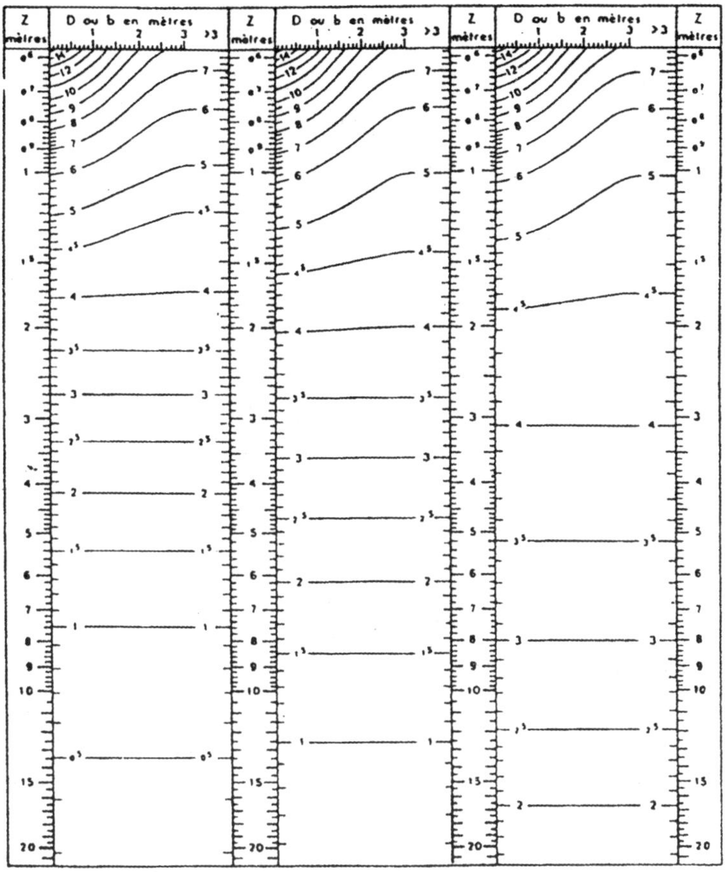

1 lane 2 lanes 3 lanes

Figure 6.21. *Static load q_o (in 10* kPa) exerted by the standard train with 25-tonne axles, at a depth Z measured from the centre line*

6.4.6. *Water pressure: presence of groundwater*

The water pressure exerted on the pipeline is caused by:

– the possible presence of groundwater,

– the circulation of a fluid inside the pipeline during commissioning.

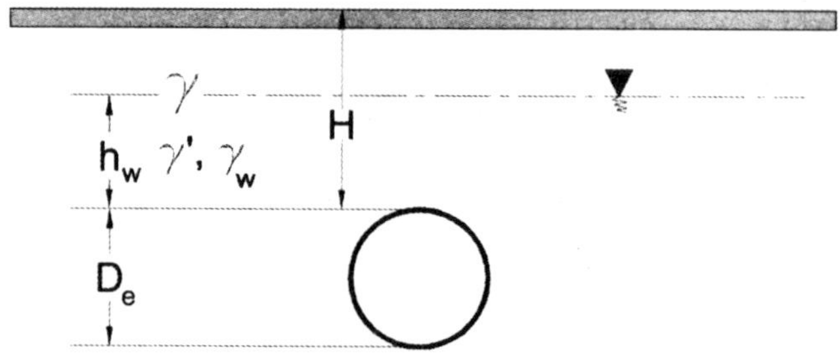

Figure 6.22. *Presence of groundwater*

Considering the Terzaghi model for pipes at low depth (paragraph 6.4.3.1), the vertical pressure due to the ground on the crown of the pipe becomes (Figure 6.22):

$$\sigma_{EV} = k_1.\gamma.(H - h_w) + k_2 .\gamma`.h_w + \gamma_w.h_w$$

with:

- γ : specific weight of the ground,
- γ : specific weight of the submerged ground,
- γ_w : specific weight of water,
- H: height of overburden over the crown of the pipe,
- hw: height of the groundwater table over the crown of the pipe.

$$k_1 = \frac{1 - e^{-2.K\tan\phi.\frac{H-h_w}{b}}}{2.K.\tan\varphi.\frac{H-h_w}{b}} \qquad \text{i.e. } k_1 = k \text{ for } h_w = 0$$

$$k_2 = \frac{1 - e^{-2.K\tan\phi.\frac{h_w}{b}}}{2.K.\tan\varphi.\frac{h_w}{b}} \qquad \text{i.e. } k_2 = 1 \text{ for } h_w = 0$$

where $b = D_e \tan(\frac{3\pi}{8} - \frac{\varphi}{4})$ and $K = \tan^2\left(\frac{\pi}{4} - \frac{\varphi}{2}\right)$.

In this way, the stress conditions which are caused by the ground, the groundwater and surface loads, at any given point P of ordinate y in relation to the pipe axis (positive downwards) become (Figure 6.23):

$$\sigma_v = k_1.\gamma.(H - h_w) + k_2.\gamma`.h_w + \gamma`.(D_e/2+y) + q_o + \gamma_w.(h_w + D_e/2+y)$$

$$\sigma_h = K.[k_1.\gamma.(H - h_w) + k_2.\gamma`.h_w + \gamma`.(D_e/2+y) + q_o] + \gamma_w.(h_w + D_e/2+y)$$

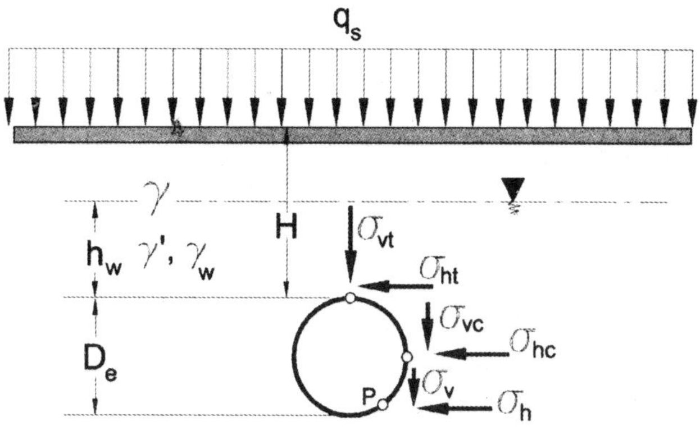

Figure 6.23. *Stresses acting on the pipe in the presence of groundwater*

i.e.:

– on the upper generatrix of the pipe:

$$\sigma_{vt} = k_1.\gamma.(H - h_w) + k_2.\gamma`.h_w + q_o + \gamma_w.h_w$$

$$\sigma_{ht} = [k_1.\gamma.(H - h_w) + k_2.\gamma`.h_w + q_o].K + \gamma_w.h_w$$

– on the pipe flanks (on the horizontal diameter):

$$\sigma_{vc} = k_1.\gamma.(H - h_w) + k_2.\gamma`.h_w + q_o + \gamma`.D_e/2 + \gamma_w.(h_w + D_e/2)$$

$$\sigma_{hc} = [k_1.\gamma.(H - h_w) + k_2.\gamma`.h_w + q_o + \gamma`.D_e/2].K + \gamma_w.(h_w + D_e/2)$$

NOTE: these calculations do not take into account the influence of the pipe's own weight.

6.4.7. *Permissible stress in the pipes*

The verification of stresses consists in comparing the maximum stress in the pipe (σ_{max}), calculated using various transverse loads, to the permissible stress, which varies depending on the materials used (see Table 6.8).

6.5. Bore fluids

6.5.1. *General information*

Resulting from the petroleum technique, the bore fluids, commonly known as "slurry" or "mud", are complex fluids that play a vital role in the construction of several sites, particularly trenchless work, whose success is largely influenced by them.

We can find a detailed description of the different functions and properties of drilling mud used for large slurry pressure TBM in the AFTES Guidelines (2002). We recall that, generally, the fluids used for boring may have several essential functions, of which some are directly related to microtunneling:

– maintaining the soil in suspension and ensuring its removal by hydraulic channels: this obviously is a function that is directly applicable to boring machines with hydraulic mucking;

– guaranteeing the stability of the bore, strengthening the walls and preventing loss of fluids by creating an external or internal cake that is as fine and as resistant as possible. This is a supporting function, which is also used in microtunneling for "lubricating" the pipe-ground interface;

– lubricating and cooling the tools, drilling strings, on-board equipment and pipelines. This last function concurs with the objective of reducing friction mentioned earlier;

– facilitating digging by jetting. This function is sometimes necessary in clayey soils (see Chapter 5).

There currently exist many products in the market; the choice of a bore fluid formulation suitable for a given site may often depend either on empiricism based on vast experience, or on several laboratory tests, which do not seem very compatible with a microtunneling site.

Air based fluids such as foams are excluded from this document because they are rarely used in microtunneling techniques. They are however the subject of research as part of the French National Research Project "Microtunnels", whose results are described in the Quebaud thesis (1996).

The AFTES Guidelines (2002) highlight the interest in establishing a "slurry program", whose objective is to optimize the choice of the type of slurry and its monitoring and inspection, in order to respond as best as possible to the technical and economic requirements of the project. We therefore need to gradually draft a summary document on the formulation of slurry, its interaction with the natural environment, its implementation, management and treatment.

We must bear in mind that, during the progress of the site, the properties of the slurry change: they depend obviously on the initial composition defined during manufacture, but also on the water and ground remains that progressively add up to the slurry: there is a need therefore to be concerned by the fluid at the start called "clean slurry" (or slurry), but also with fluids polluted with excavation remains ("polluted slurry" or mud).

The following are the main characteristic parameters of drilling mud, which determine its behavior and which must be regularly measured and recorded as the work progresses:

– the density, which is an index of the content of solid element in the mud; it must generally be between 1.0 and 1.2;

– the viscosity, which characterizes the ability of forming a cake as well as the ease in transportation of the mucking; measured at the Marsch cone, it must generally be between 32 and 40 seconds in clayey soils, and greater than 50 seconds in sandy soils;

– the yield point, the thixotropy and the filtrate that determine the formation of the cake and its ability to reform rather rapidly; in a filtration test, clean slurry must present a cake less than 4 mm and a filtrate less than 40 cm^3; in mud, the cake must remain less than 3 mm, and the filtrate of the order of 6 cm^3 in clayey ground, and 10 to 15 cm^3 in sandy ground;

– the sand content, which results from the separation result of solid earth and which affects the permeability of the cake and therefore its stability; it must generally remain less than 4 to 5 % (measured with the elutriator);

– the pH, which affects the ionic balance and thus the physico-chemical properties of the slurry; it must remain with a range of 8 to 10;

– the conductivity and the hardness are also indices that may be useful.

NOTE: the values indicated above are only to indicate the order of magnitude generally used in trenchless work. They must, however, be adjusted in accordance with predominant performances, which are obtained from the ground. We will give some indications on the expected performances in paragraph 6.5.2.

We can also modify during the construction at the site the properties of the polluted slurry, in order to restore the desired properties which may have progressively deteriorated: according to the rheological and filtration properties required, a simple suitable physical or chemical treatment may be carried out.

Finally, the residual sludge, i.e. slurry that cannot be reused for the site, will have to be treated or eliminated; this operation has become very restrictive due to the change in legislation. It will be dealt with in paragraph 6.5.6.

6.5.2. *Selection criteria*

The level of characteristics required will depend on the difficulties of the project: geometry, complex and varied geology, unfavorable geotechnical elements, abnormalities, pollution of ground. The greater the number of difficulties, the more precisely will the minimum required criteria have to be specified and analyzed before and during the work.

A priori, the selection of the type of slurry is done by answering the following main questions:

– will the slurry be recycled or not? This often depends on the estimated slurry volume, the equipment used;

– what are the main functions desired: stability, mucking, lubrication, ease of removal, etc.?

– what are the quality criteria for the composition of the slurry?

– what is the degree of complexity of the implementation and checks to be carried out at the site?

In practice, this selection is often dependent on:

– nature of the project and levels and quality of information available (study of the ground, etc.);

– the equipment, experience and the competence intrinsic to each company.

Table 6.10 (FSTT RT 30) specifies the impacts of different classes of soils (classified according to NF P 11-300 standard "Classification of materials that can be used in the construction of backfills and layers in road foundations"), and therefore the main functions required from the fluid.

Class of ground	Standardized classification of grounds	Impacts on the microtunneling project
A	Fine soil: clayey silt	Sticking, tamping with possible jamming
B	Sandy and gravely soil With fines: sand with gravel More or less clayey	Abrasiveness, deviations, and possible implosion followed by jamming
C	Soil including fines and large elements: clay or flint chalk, grinding grit, talus scree, moraines	Abrasiveness, deviations, and possible implosion followed by jamming
D	Soil insensitive to water: Sand with clean gravel	Partial to total loss of mud Instability of the walls
R	Rocks: carbonated, clayey, siliceous, saline, igneous and metamorphic	– Rapid wear of tools and shafts – Contamination by salts (evaporates) – Difficulties in directing the drilling in soil-rock interfaces – Loss of mud and instability in fractured areas
Special Materials	Organic soil, industrial sub	Mud loss, contamination by organic matter

Table 6.10. *Impact of different types of soils on the drilling mud and supervision of the site*

6.5.3. *Products used*

The drilling slurry is essentially made up of a stable colloidal suspension in a dispersing agent: water. Two families of colloids are mainly used:

– minerals: mainly bentonites,

– organic: mainly water-soluble polymers.

This suspension is rapidly altered by solids in the ground and possibly by water contained in the ground to be crossed and the minerals contained in it.

The bentonites are industrial clay of the smectite group. They are characterized by a foliated structure, which is negatively charged on surfaces and positively on fractures. Upon contact with water, the flakes disperse, swell and possibly exchange the charge compensating cations. Beyond a certain concentration (relatively low, of the order of 4 to 6%), and depending on the quality of bentonite, a stable structure develops and has a certain rigidity under shearing.

The bentonites can be combined with *additives* for various functions: viscosifying, fluid-loss additive, water reducer, clay encapsulator and stabilizer, lubricant. The most common are water-soluble polymers, which in addition to their ability in increasing the viscosity present special physico-chemical properties. There exist several types that are natural, artificial or synthetic, which can remedy specific problems relating to certain soils, such as:

- sticky or swelling clay,
- improvement in the stability in sand and gravel,
- better resistance to physical or chemical contaminations, etc.

The family of products generally used in the market for trenchless work are:

Bentonites	Polymers	Others
Bentonites with high efficiency	Viscosifying polymer of type PHPA or cellulosic	Surface-active lubricant or detergent

In practice, companies that have not defined their own standards for boring fluids, may refer to the NF EN ISO 13500 standard "Boring fluids: specifications and tests" dated September 1998.

The manufacturers provide a certain number of documents that enable the use of the products under trade practices: technical and safety data sheet (NF ISO 11014-1 standards "Safety data sheets for chemical products" dated November 1994), technical notes and application sheets.

6.5.4. *Recycling and processing*

Depending on the volume of the slurry to be processed, a site may decide to work only with clean slurry during the digging stage, or to provide for a treatment and recycling station for boring fluids. In the latter case, the important steps are: grit removal, desilting and hydrocycloning.

In practice, the earth is separated with the help of vibrating screens, which separate the sand then the silt from the fines, then the cyclones that remove the finest elements by centrifugation. We get a coarse fraction, which can be reused, and a "pulp" made up of the finest elements. However, the active clay content (from bentonite) must be regularly monitored, as the slurry gets mixed with "inert" fines from the excavated ground, which are not extracted by hydrocycloning and alter the properties of the slurry.

The recycled slurry must therefore be regularly reclaimed, because of this gradual loading of inert fines, but also by contaminations caused by groundwater, and due to the consumption of bentonite and water by filtration in the ground. Different types of treatments can help soften the recycled slurry:

– adding new slurry before sending it back in the drilling system,

– adding additives (viscofiers, plasticisers) to correct the characteristics that have become non-compliant.

6.5.5. *Implementation at the site*

Implementation of drilling mud requires suitable manufacturing, storage and solid treatment equipment:

– mixers,

– main and auxiliary pumps,

– mud tanks, vibrating screens, hydrocyclones, centrifuges, and possibly a plant for the physico-chemical treatment of waste.

The equipment, as well as the quality of process water and the temperature, will significantly affect the performances of the slurry. On the other hand, we must also point out that the storage conditions (long periods, humid atmosphere, etc.) can significantly alter the characteristics of the bentonite powder.

6.5.6. *Slurry treatment: technical and regulatory aspects*

6.5.6.1. *General considerations*

The purpose of this chapter is to define technical and regulatory criteria for the management of drilling waste according to the regulations in force, and to present the possible lines of treatment for drilling mud. It includes the main conclusions of the French National Research Project "Microtunnels" (FSTT RT 27). In fact, consideration of the treatment of slurry is decisive, in terms of respect for the environment as well as in financial terms; this aspect has become essential.

Slurry from treatment plants that cannot be used in that condition anymore is compared to waste that must be treated in order to be stored. This waste is either inert or dangerous depending on the cases considered. Nevertheless, like all waste, slurry belongs to the producer and therefore to the boring company and it must be treated before finally being suitably stockpiled (according to law no. 75-633 dated 15th July 1975, modified by law no. 88-1261 dated 30th December 1998, law no. 92-646 dated 13th July 1992 and law no. 95-101 dated 2nd February 1995).

Not too long ago, the largely polluted effluents were discharged into the natural environment without any special precautions. If the purification capacity of the environment (very often the water courses, or abandonment at the site) was limited, malfunctioning occurred. This local pollution was in part controlled and the development of pollution removal techniques transferred this essentially liquid form of pollution towards the solid waste sector. Naturally, this type of waste has become illegal, and can lead to legal proceedings.

Currently, drilling stations are equipped either with plants where the fluids are essentially dewatered before disposal on land of various sub-products, or storage units where the residual fluids are provisionally stored before being collected by trucks and taken for disposal, or treatment plants. Removing water, which is the main enemy of disposal due to the generation of leachates, remains the prime objective. The waste storers that do not themselves carry out the operations of elimination or reclamation must call upon a private or public collecting body. In case the waste is abandoned, the incumbent government may undertake this elimination at the expense of the manager.

In addition, the producer is interested in constituting a reference file, in order to ensure the traceability of products, which must include the information necessary for the identification of the waste produced, and some overall analysis results on representative samples. However, a microtunneling site with its treatment tool (treatment and recycling unit) differs from other treatment plants by its provisional and mobile nature.

Considering the nature of the sites and the diversity of products, our analysis has led us to define three criteria to be considered by the company to manage wastes generated by excavation work and to define a suitable treatment line (see Figure 6.24):

– *site criterion*: in addition to the geological conditions: polluted site identified, polluted site not identified, site known as non-polluted;

– *quantity criterion*: site duration, volume to be treatment, treatment location. The volume of slurry to be treated varies from a few m^3 to several dozen m^3 per day;

– *slurry criterion*: mineral slurry, organic polymer slurry, mixed slurry.

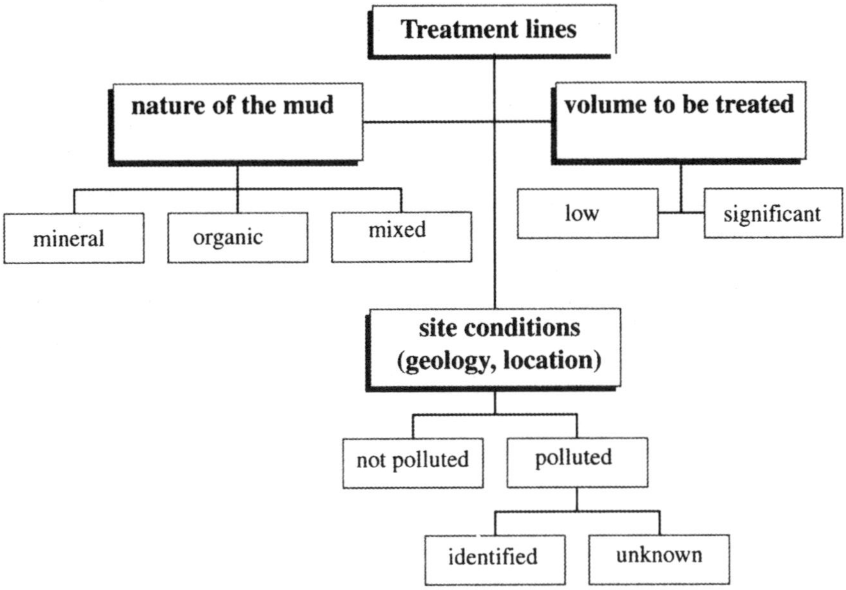

Figure 6.24. *Definition criteria for treatment lines*

Figure 6.25 summaries the successive stages of the treatment chain.

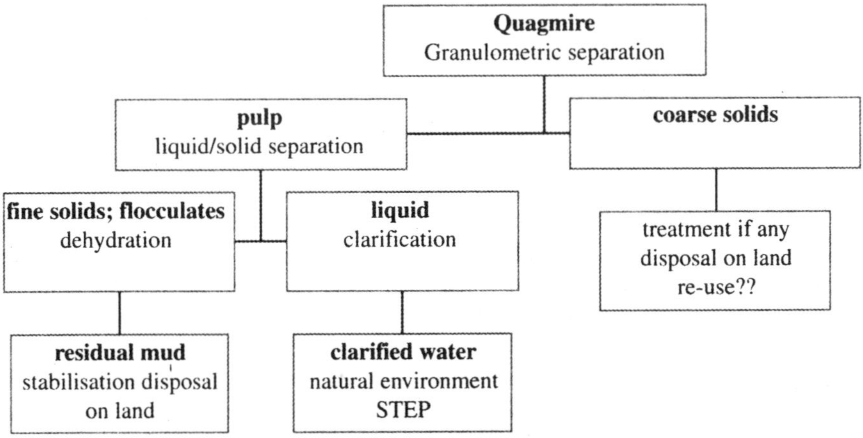

Figure 6.25. *Logic diagram of the main stages of treatment*

6.5.6.2. *Current regulations*

We present in Appendix 2 a summary of the main texts in force on waste regulations.

6.5.6.3. *Lines for removal of drilling residues*

Depending on the waste legislation, the drilling mud and other waste comparable to drilling mud will have to be classified in relation to the source of the bore fluids used and the characteristics of the ground crossed (hydrocarbon, heavy metals and organic matter content). *The soil criterion is fundamental here*, as inert waste mixed with dangerous waste becomes dangerous.

In spite of their essentially inert nature, particularly if the excavated site is identified as non-polluted, the bore fluids cannot be discharged into the natural environment, because their abandonment causes damage to the site or pollution to watercourses by large amounts of sediment loads, which can have a harmful effect on the soil, the flora and the fauna. Several solutions or destinations for the discharge of waste mud, containing soil material, are possible:

– removal without prior treatment: to treatment plant or as discharge;

– treatment at the site with development of a treatment tool at every plant;

– recourse to a collection system within the company or sub-contracted to a specialized treatment centre.

The last two solutions use the same processing techniques (sieving and liquid/solid separation). They are selected according to a criterion for localising the site (urban or remote), and a criterion depending on the quantity of mud produced during excavation work.

6.5.6.3.1. Discharge without treatment

Direct disposal into the public water system to the treatment plant of the concerned town may be possible if the quantities are small. The sewer systems form part of sanitation installations of communities. They discharge various urban effluents thrown out by individual, certain commercial, craft and industrial activities to treatment plants. The inflow of such mud may, however, disturb the complex chain for the elimination of pollution.

This solution is discussed on a case by case basis with the town council or the water treatment plant manager, but after several consultations and considering the composition of mud (too much mineral matter, no organic matter and some metals), we have been advised against employing it by operators, as these discharges may cause risks to the proper operation of their plant.

Direct disposal on land without treatment is sometimes used; however, this solution is far from satisfactory.

6.5.6.3.2. Treatment lines

Generally, the companies must therefore either carry out treatment at the site (liquid – solid separation unit), or send the waste to a temporary storage centre where the mud from various sites is collected, or send it to a processing plant adapted to handle large volumes (this process is particularly profitable in urban areas).

Moreover, if the company decides to develop its own treatment tool, we must then identify whether the site belongs or not to the nomenclature of designated plants. If it is listed, it must be subject either to notification, or authorization and comply with the order of 2nd February 1998. On the other hand, if it is not so, it must then be subject to the decree dated 29th March 1993 of the Waters Act.

Solid parts (coarse and fine) of the mud

Currently, the drillers negotiate the storage condition of solids. Disposal on land remains the most viable and the most commonly used methods. To be disposed on land, it is necessary that the solids have a dryness of at least 30%. Depending on its nature, the waste could be disposed on land as class 1, 2 or 3:

Class 3 (as a general rule): the waste under this class is produced by microtunneling and horizontal drilling operations in non-polluted soil, identified or verified, with the help of bore fluid: water, water/bentonite mixture, without the addition of additives in large quantities.

Class 2 (occasionally): concerning the drilling mud, this category includes mud that is mixed with large quantities of polymer additives. The theoretical evaluation of the DOC of a suspension of 0.1% of CMC (1 g/l) gives a value of 740 mg/l. This mud may be considered as being equivalent to fermentive and rapidly changing waste of the industry. This aspect remains to be dealt with in depth.

Class 1 (exceptionally): the drilling mud coming under this class is:

– drilling mud with hydrocarbon content of more than 1%,

– boring residues resulting from the use of boring fluids with low hydrocarbon content,

– residues from the treatment of polluted soil.

Some polluted soils may be classified as inert if the results of the polluting potential tests, which include three successive leaching according to the standard NF X 31-210, do not show any release of pollutants in the leachate.

Effluents or clarified water

This water can be discharged either directly in the natural environment if the discharge standards and the receiving body of water permit it, or in the public water system towards the STEP (provided that the standards are respected), or be regarded as a leachate if it complies with the standard in force.

Storage in reservoirs or pits must comply with the legislative provisions in force. It is the same for decanting systems and transportation to treatment plants.

It must be recalled that in the case of water sampling to prepare drilling fluids (either from a unconfined groundwater, or from a watercourse), the company must restrict its water consumption and equip the sampling stations with measurement devices.

6.5.6.4. *Prospects for reclamation*

Reclamation can be envisaged only for waste generated during excavation work done with mineral slurry containing very few additives, in soil not having or having very little polluting potential and if the operation is economically viable and technically possible.

If all these conditions are satisfied, the earth of the drilling mud can be reclaimed at their production site itself or after transfer to a platform equipped to gather and process all the waste from a production, equivalent to the size of a town.

Some possible methods for reclamation in the field of civil engineering may be suggested:

– soil mortar ready for use,

– clean material ready to be used for pavement structures,

– sieve correction of soils; particularly mud containing bentonite could help make some soils watertight if this function is sought,

– use of ultra fine particles in the concrete.

These techniques are still not well developed, and do not have any immediate application that can be envisaged for waste from microtunneling sites.

Chapter 7

Guidelines for the Site Supervision

7.1. Guidelines for guidance

7.1.1. *Necessity of controlling trajectory deviations*

It is essential for the success of a microtunneling project to follow the trajectory and minimize the deviations.

Firstly, when creating a gravity system, it is essential to fix a well-defined angle.

In addition, trajectory deviations brought about by the misalignment of the pipes and by the fact that the drive thrust is not centered on the axis of the pipes and the distribution of stresses is no more uniform on the section. For the same jacking thrust, the maximum stress applied on the section of the pipes will therefore increase. When calculating the acceptable thrust on the pipes, an angulation of 0.5° between the pipes is generally taken into account by the manufacturers (see section 6.3). Thus it is necessary to minimize the angulations between pipes by ensuring continuous trajectory corrections.

Finally, sharp increases in thrust at the head can occur due to trajectory corrections; the contact stresses on the body of the machine during changes in direction become more significant (see paragraph 3.1.3.5).

Furthermore, it is noted that during stable excavations, the horizontal trajectory deviations were caused by an average increase of 50% in frictional forces (see paragraph 6.2.4.2.2). According to the studies carried out by Milligan and Norris

(1995), the amplitude of these increases in friction is linked to the curvature of horizontal deviations.

7.1.2. *Guidelines for the measurement of deviations*

The position of the machine is marked according to the planned course. The parameters measured are: vertical deviation (EV), horizontal deviation (EH), the angle (IV) and the azimuth (IH) of the axis of the machine with respect to the theoretical axis. The operator modifies the trajectory based on this data; it is thus essential that these measurements are precise and reliable.

Vertical and horizontal deviations are obtained using a laser beam. The laser is fixed in the starting shaft. It emits a beam in the direction set beforehand according to the desired trajectory. This beam is reflected onto a target located at the back of the machine body.

The laser must be fixed to a part of the shaft, which is not attached to the dead man. In fact, the latter supports the thrust frame, which can be subjected to movements during the implementation of the thrust cylinders. Even a slight modification of the inclination of the laser stand can cause significant differences at larger distances in the deviation readings. These deviations naturally increase with the length of the shaft; furthermore, the increase in drive thrusts at the end of the section can lead to significant movements of the thrust *massif.*

Thus, experimental data has shown that from 80 m onwards guidance problems inherent to these measurement inaccuracies appear: the amplitude of most of the deviations at the end of a section are between 10 and 30 mm, whereas half of them are less than 10 mm before 80 m.

Thus it is advisable to fix the laser stand at the base slab, which is less likely to move than the walls of the shaft on which the dead man is attached. Another alternative would be to fix it on a support attached to the surface of the ground and independent from the walls of the shafts.

Some lasers have an automatic vertical leveling based on the concept of a plumb line, thus compensating for the vertical movement of the laser stand. As regards its positioning in the horizontal plane, a continuous adjustment is required.

7.1.3. *Guidelines for the monitoring of deviations*

Experimental data helped provide some explanations on the problems of trajectory deviations, which could be noted and thus draw up recommendations for the different phases of a site: initial adjustments, corrections during jacking, dimension of the overcut.

7.1.3.1. *Initial adjustments and starting of jacking*

If the thrust system is not positioned correctly or if an initial misalignment occurs, significant corrections will be required to restore the trajectory. Experimental data of the National Project has shown that major deviations are located within the first 20 meters. A large proportion of trajectory deviations (30 to 50%) are between 30 and 40 mm (see paragraph 3.1.2), whereas between 20 m and 80 m, the majority of deviations do not exceed 20 mm. To check the initial deviations we must ensure that:

– the thrust frame is correctly positioned and oriented according to the trajectory fixed for the pipes. Correct positioning of the base slab facilitates the leveling of the thrust frame;

– the frame does not lean under the weight of the boring machine and the ground entry ring is correctly aligned on the trajectory and does not lead to an initial misalignment of the pipes.

The difficulties caused by reworking of the ground near the walls of the shaft to retain a correct slope and direction of the boring machine can also be emphasized. As a result, the entry of the boring machine into the ground is a tricky phase that requires frequent monitoring and adjustments. During this phase it is advised that the boring machine be guided using its body rather than its head rams.

7.1.3.2. *Corrections during jacking*

The National Microtunnels Project has shown that a sudden correction in the trajectory often led to an increase in deviations. The following recommendations can therefore be suggested:

– to minimize the amplitude of deviations, the operator should anticipate his action on the steering cylinders by taking into account the response time of the machine. In fact, as shown in paragraph 3.1.2.1, the body of the machine follows the angle imposed on the head by the steering cylinders only after several meters of jacking (an average of 4 m). While covering this distance, the boring machine continues to follow the previous direction, thereby creating a new trajectory deviation in the opposite direction (see Figure 7.1);

– successive small corrections should be preferred even though they are slow in restoring the theoretical trajectory rather than sudden corrections, which can lead to small radii of curvature and thus cause significant angulations between pipes (see Figure 7.1). More than the amplitude of the deviations, it is the angulations between pipes, which cause most problems (see paragraph 7.1.1). It is advised that an angulation of 0.5° not be exceeded (see section 6.4). Furthermore, with small corrections even the amplitude of over-corrections mentioned earlier is reduced.

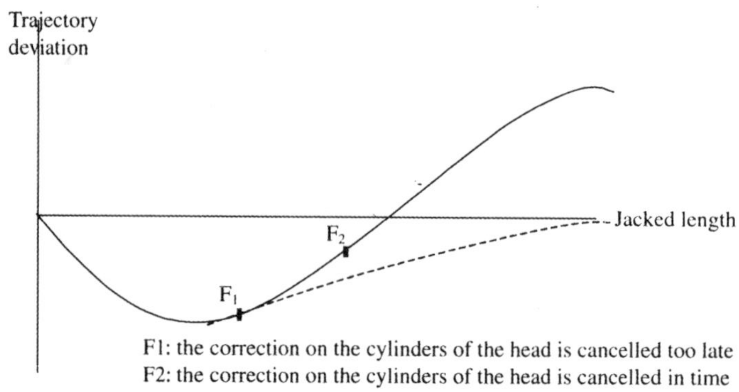

Figure 7.1. *Trajectory followed by the boring machine according to the anticipation of corrections*

It is to be noted that the PJA (*Pipe Jacking Association*) provides assistance in decision-making for manual control. The objective is to determine the possible corrections on the head rams by fixing a limit for the angulation between pipes (for example $\alpha = 0.1°$). There is also software that helps automate the corrections made to the boring machine (Phelipot, 2000). Some are based on the concept of fuzzy logic, taking into account at each step the response of the machine during previous stages. Other software (Iseki and Decon-Soltau) relies on a database collected from different soils. However, according to Anheuser, automatic guidance is not necessarily the best solution (Anheuser, 1994).

7.1.3.3. *Adjustment of the overcut*

The value of the overcut also comes into play in facilitating the steering of the boring machine and thus reducing the deviations. A minimum overcut is in fact required to guide the boring machine, but it should not be too large, as it would make the corrections on the head cylinders less effective, the machine having no sufficient support on the walls of the excavation. In France, for boring machines having diameters between 600 and 1,200 mm, an overcut (difference in radius between the cutterhead and the pipes) of 20 to 30 mm is generally used.

7.2. Guidelines on the drilling parameters

The drilling parameters known by the operator are generally: the total jacking thrust, the engine torque driving the cutterhead, the mucking flow rate, the jacking speed, the roll, the position in the plane and in altimetry.

It can intervene preferentially on the first two parameters so as to monitor the excavation of the soil and the transportation of earth and thus act on the on the value of the thrust at the head. It is important to monitor the thrust at the head at several instances:

– it should remain greater than the active pressure of the soil so as not to generate significant compaction, and lower than the passive pressure of the soil so as not to cause an upheaval of the ground (see paragraph 3.2.3);

– control over the thrust at the head can help significantly reduce the total thrust and thus increase the jacking length between two shafts.

Unfortunately, the boring machines are currently not equipped to measure the thrust at the head. Only a parameter characterizing the torque of the wheel is measured. This depends on the way in which the wheel manages to cut and extract material from the face and the difficulty in crushing and evacuating it through the mucking pipes. The adjustment of the mucking flow rate and the jacking speed will therefore be done according to the value of the torque.

7.2.1. *Avoid instability of the face*

When jacking is done above the water table, the contact pressure of the cutting head is sufficient to balance the ground pressure due to the small excavation diameters.

In the case of an insufficient contact pressure, losses in bore fluids in the soil can also lead to instability of the soil. The objective is to avoid excessive erosion at the face, which will lead to an excess removal of soil in relation to the penetration of the machine and possibly (if the soil is permeable) to losses in mucking fluids in the ground. For this, it is necessary to maintain the torque above a minimum value generally indicated by the supplier of the machine.

To remain above the minimum torque value, the operator can:

– either increase the jacking speed,

– or reduce the mucking flow rate; in the case of coarse and permeable soil, the use of high density slurry helps maintain lower mucking flow rates while ensuring the transportation of crushed stones.

In addition, to reduce the losses of mucking fluid in the soil, it is better to inject the mucking fluid into the mucking chamber and not into the crushing cone. The opening and closing of different valves is generally controlled from the operating station.

7.2.2. *Avoid excessive thrust on the head and the blocking of the cutterhead*

The thrust at the head can increase considerably when the rate of extraction and discharge of muck is low in relation to the jacking speed applied. The jacking is then accompanied by a discharge of the soil.

Microtunneling can encounter these problems in clayey soil. In fact, the discharge of muck in such soils can pose some difficulties linked to the sticking of material leading to the clogging of pipes and the crushing cone.

To avoid excessive thrust the operator may:

– reduce the jacking speed,

– set high mucking flow rates,

– reverse the direction of flow of the muck at the level of the boring machine using a bypass system, the temporary isolation helping moreover to rinse the pipes,

– inject water at high pressure at the cutterhead to clean the tools of the head and the crusher,

– inject the mucking fluid in the crushing cone and not in the mucking chamber so as to facilitate the passage of excavated earth from the crusher to the mucking chamber through the openings located on the lower part of the cone.

It is important to monitor the quality of the mucking fluid and renew it if necessary.

In permeable soil, a faulty return of earth can also be caused due to losses in mucking fluids by seepage into surrounding soil or by its return into the starting shaft through the annular space. By increasing the viscosity of the mucking fluid, these types of risks can be reduced. Furthermore, for coarse soil, a sufficient flow rate is necessary to transport large elements; in this case the use of high viscosity mud is advised to enable lower flow rates.

7.2.3. *Checking the roll*

In order to control the roll in the best possible way, i.e. the rotation of the entire boring machine with respect to its axis, it is convenient to ensure a sufficient contact between the skirt of the machine and the ground as well as to reduce the torque on the cutterhead. These requirements are represented by the following recommendations:

– do not inject lubricants in the first few meters of jacking, the injection nozzles being situated at the back of the machine;

– check the torque at the head, a strong resistance on the drilling wheel can lead to a rotation of the boring machine. On the contrary, in the case of permeable soil it is better to maintain the torque at a value higher than the minimum so as to avoid losses in mucking fluids: this, on returning from the overcut, can excessively "lubricate" the machine and can be detrimental to the control of the roll.

7.3. Guidelines on lubrication

All boring machines are equipped with the ability to inject a lubricating fluid into the annular space created by the overcut. The injection of the lubricant is particularly recommended; it considerably reduces the dynamic frictional forces and limits the increase in thrust after a stoppage.

Paragraphs 3.1.3.3 and 6.2.4 have shown that the decrease in frictional forces caused by lubrication can vary from 45 to 90% depending on the injected volume, the lubrication system and the type of lubricant used. In view of these observations we can emphasize the following facts:

– to make the injection of the lubricant effective, an overcut must be made so that the lubricating fluid can flow between the soil and the pipeline. The width of the overcut must be chosen depending on the possible risks of swelling of the clay, the amplitude of flexible dumping of soil and should surely not cause problems of settlement;

– an injection of the lubricant at various places along the string of pipes and not only at the back of the machine ensures a good lubrication of the jacked section;

– it is better to lubricate continuously even if the thrust stress is low. Losses in efficiency and an improper control over the frictional forces is noted when the injection is done discontinuously, as a reaction to the increase in stresses. The injection of fluids in the annular space is also useful in opposing the convergence of soil on the pipes during stoppage;

– the volume of lubricant injected must at least be equal to the volume left by the overcut. Greater volumes can help stabilize the excavation, thus reducing the frictional forces created only by the weight of the jacked pipes (see paragraph 3.1.3.4 and section 6.2). It is therefore important to be able to quantify and monitor the volume of lubricants injected;

– the pressure of the lubricant must also be measured and monitored. A high pressure can lead to a rise of the fluid to the surface.

The influence of the characteristics (density, viscosity) and the nature of the lubricating mud (bentonite, polymers, mixed mud) are discussed in section 6.3. Up until now, the quality and nature of the mud were generally not determined according to the nature of the soil at the sites. It will however be necessary to specify the chemical and physical interactions of the lubricating mud with water and soil with the help of additional studies.

For example, it seems vital to take into account, for the choice of the lubricant and the injection rate, the fluid looses due to seepage into the permeable soil (see paragraph 3.1.3.4).

In addition, care must be taken in the case of swelling clay so that the water content in the lubricating fluid does not itself become an aggravating factor in swelling by making a judicious choice of the fluid (see section 6.3).

7.4. Guidelines regarding stoppages during jacking

Stoppages during jacking can lead to an increase in thrust at the time of restarting caused by the creep of soil into the pipes and by the dissipation of excess pore pressures of the lubricating fluid (see paragraph 3.1.3.3). After certain drilled distance, the restarting thrusts being the most penalizing, it is important to be able to predict and monitor them.

7.4.1. *Provision for the increase in the thrust during restarting*

The amplitude of additional friction caused during the start is first of all linked to the stoppage duration; a semi-logarithmic relationship was demonstrated between the increase in thrust during restarting in relation to the thrust recorded before the stoppage (DPx100/P), and the logarithm of the stoppage duration (in hours) (see Figure 7.2).

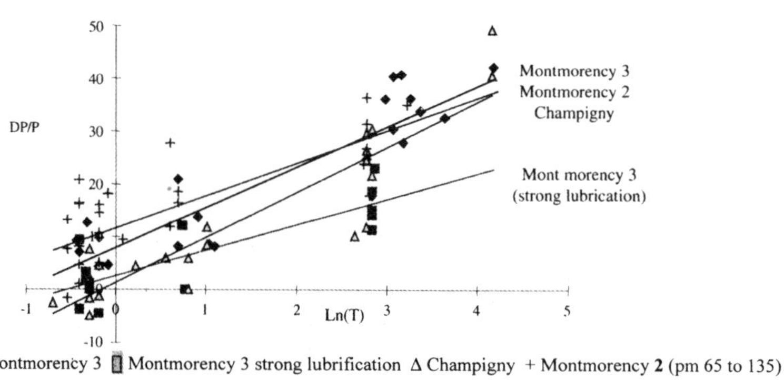

Figure 7.2. *Rate of increase in the total thrust according to the duration of interruption in jacking*

Due to the large scattering of the points, particularly for short interruption durations, it is difficult to define a general relation. It must be noted that, on average, there is an increase in the total jacking thrust by 25 to 30% after an interruption of one night and an increase of about 40% following a weekend stoppage. These values depend on the nature of the soil and the lubrication conditions. It is better to extrapolate during the project from the values obtained for stoppages of a couple hours and one night to estimate the consequences of a stoppage of several days and to check the value of additional friction taken into consideration during dimensioning (see section 6.2).

7.4.2. *Limit the increase of the thrust during restarting*

It seems surely preferable to manage the progress and the planning of a site so as to avoid long interruptions, particularly at the end of the section when the total jacking thrust reaches values such that an increase of 40 to 50% can lead to thrusts exceeding levels acceptable by the pipes and the capacity of the thrust cylinders.

A constant and sufficient quantity of lubricant injected during jacking (see paragraphs 3.1.3.3 and 3.1.3.4) helps reduce the additional friction following interruptions in jacking (see Figure 3.11). The fluid injected between the soil and the pipes can in fact work as a retaining structure and stabilize the excavation if its volume is sufficient to create a watertight cake at the soil-fluid interface and fill up the entire annular space.

7.5. Data acquisition during the project

According to the EN 12889 standards of May 2000, the following data must be recorded if possible automatically during the project and retained:

- position in the plane and in altimetry,
- guidance corrections,
- maximum drive thrust from the main thrust bench, and thrust stations if need be,
- speed and distance of penetration,
- quantity of lubricant and, if possible, the soil excavated,
- roll.

The standard recommends a maximum recording interval of 0.2 m.

In addition, it is recommended to record the settlements and upheaval of soil generated during jacking.

Chapter 8

Socio-Economic and Contractual Aspects

The National Research Project "Microtunnels" has helped undertake several investigations to improve the different technologies used in the microtunneling market. However, even if technological development plays a predominant role in the approach of companies, it can only expect a development if socio-economic considerations are not taken into account in the same capacity as the technical priorities.

This is why, in the framework of these guidelines, it seemed important to specify, on the one hand the socio-economic context of urban sites and on the other the contractual aspects that determine the microtunneling market.

8.1. Social and economic aspects: concept of social cost

The inhabitants of a town or district do not appreciate the repeated opening of trenches on their roads. The temporary closure of a pavement for work can give rise to protests by shopkeepers and can sometimes endanger the safety of pedestrians. The reduction of traffic corridors can cause traffic congestion and traffic jams. In brief, trenches are undesirable.

All network administrators are faced with this problem. The intensification of surface traffic, increasing traffic congestion on the underground and requirements of the communities make the use of new operation techniques on networks inevitable.

The problems of urban sites are currently well known; they, nevertheless, do not form part of the calculation of total costs of a project and do not figure in any regulations that attempt to reduce them. On the other hand, this type of work lacks efficiency in carrying out evaluations, comparisons and annual estimates so as to deal with these problems in some other way.

The objective of this chapter is to recall the socio-economic conditions of urban sites and take decisions to justify or not the use of innovative techniques for the laying or renovation of underground pipelines.

8.1.1. *Value of modern urban sites*

8.1.1.1. *Total cost of the work*

A recent report, prepared in 1997 by the Central Laboratory of Bridges and Tunnels (LCPC) at Nantes and the INSITUFORM group France, defines the concept of total cost of a new or renovated structure:

Total cost = Direct cost + Overhead cost + Social cost

8.1.1.2. *Direct cost*

Direct cost represents the total amount required to build the structure. In a context where the direct cost is considered as the only criteria for the calculation of the estimate of a project, stress is laid on an objective consideration of the nature of soil, problems of reinforcements, depth, obstructions and the presence of water.

In this way, Michelizza (1991) showed the economic interest of boring machines compared with trenches: thus, for laying a 600 mm diameter pipeline at a depth of 6 m, the boring machine proved to be competitive from 30 ml length in comparison with a traditional tunnel and a jacking of 1000 mm (minimum diameter required by the technique). Among the published costs, it is also interesting to state the study carried out by the *Transport and Road Research Laboratory* (Norgroves and O'Reilly, 1991), based on the costs of 16 offers and 11 actual projects for the laying of sanitary drainage pipelines using trenches and/or microtunnels between 1970 and 1981.

8.1.1.3. *Overhead cost*

The overhead or induced cost represents the total amount of additional work on the project, very often prior to the execution of the structure (for example, contract deviations, moving of street furniture, setting up of traffic signals for deviations). This cost is not always accounted for in the microtunneling market and must form part of a monetary quantification integrated into the cost of the project.

8.1.1.4. *Social cost*

In 1978, Beauvais defined "*the social cost of an activity as all the inconveniences faced by the society due to this activity*". Obviously, the market does not sanction the cost and its monetary quantification is not easy.

In order to understand the soçial cost of a project and thereby estimate the total cost, it would be better to bear in mind the nuisance factors relating to an urban site and to specify methods for estimating the social cost.

8.1.2. *Traditional urban sites: nuisance factors*

The installation, renovation and maintenance works of underground urban technical networks cause problems of integration within urban areas. Every year, the digging of trenches in urban areas causes significant disruption to traffic, adjoining businesses and the lifestyles of residents.

Even if experiences show faults and difficulties in controlling the inconvenience caused by the work, efforts were, however, made to attempt to diagnose the nuisances as the work at the site progressed, with the help of investigations carried out by the university survey – only one in 1991 (Angot) – or through discontinued studies such as the study by the Ingénieurs des Villes de France in 1993 (Vidal). The different studies have highlighted four major categories of nuisances:

– traffic inconvenience,

– environmental degradation,

– accident hazards,

– economic impacts.

8.1.2.1. *Traffic disruption*

The construction of a site on the traffic system can have an impact on all the movement in and around the area:

– the movement of site equipment disrupts the traffic,

– detours and delays caused by the works can partially or totally block the traffic ways,

– access to public or private resorts is reduced (inconvenience to pedestrians, deliveries, mail, collection of garbage, etc.),

– parking facilities are reduced (the site occupies parking space, an alternative route blocks the parking area, etc.),

– public transport is diverted or disrupted. Additional buses are required to maintain the frequency of services.

8.1.2.2. *Damage to the environment*

The impact on the environment by an open trench site affects firstly the surroundings of the site:

– the noise created by the works is often unpleasant (pneumatic drills, electric generators, power shovels, etc.). The vibrations caused by the equipment on the site can have physical and psychological effects on individuals;

– air pollution by exhaust gases is increased due to traffic slowed down by the site (emission of CO, CO_2, hydrocarbons, lead, nitric oxides, etc.). Unpleasant smells emitted from the trenches during works in sewers are added to the emission of exhaust gases by the equipment at the site;

– dust and dirt appear following excavation of backfills. This may force the residents to carry out additional cleaning of their cars and windows. Near more sensitive areas such as hospitals or laboratories, the particles carried by the wind can have more serious effects;

– the disorder and visual impact of the site are unpleasant, particularly on pedestrian streets, at tourist sites or near historical monuments;

– landscaped areas may be damaged and cutting down trees becomes inevitable at times.

Furthermore, an important part of the nuisance is scattered at the local, regional or even national level:

– the trucks that transport the backfill and rubble, sometimes over a distance of tens of kilometers, block the roads, create noise and pollute the air all along their way. Heavy goods vehicles are responsible for approximately 50% of the polluting emissions by traffic;

– the dumping area where the rubble removed from the trenches is dumped, as well as the quarry from where the natural soil is drawn for the backfills, are also nuisances for the surroundings of the residents;

– the transportation of backfill and rubble by trucks consumes large quantities of non-renewable energy.

8.1.2.3. *Risk of accidents*

An open trench site can cause numerous accident hazards for the workers at the site as well as its environment:

– the workers at the site are exposed to falls, collapse of the trench and to the dangers during the handling of pipes;

– if built under poor safety conditions (very narrow or no protected cross walk, no signaling of work, operation of equipment, etc.), an urban site can be a public hazard;

– for motorists, in the case of partial roadblocks, the risk of accidents can increase (temporary road works, regulated by the traffic lights or by a worker);

– in addition to the hazards at the site itself when the traffic is diverted, the danger for the residents' alternate route is indirectly increased. A sudden increase in the traffic can endanger the lives of children and the elderly. Furthermore, motorists are forced to drive on a road not designed for such dense traffic.

8.1.2.4. *Economic impacts*

An open trench site can have a strong economic impact on the surroundings in the urban context:

– the site can be responsible for a decline in commercial activities. If customers cannot reach the shops, shopkeepers will register lower sales. Government revenues (taxes, duties etc.) will be reduced in the same proportion. More seriously, long-term or frequent projects can affect the economic vitality of the area and discourage possible investors;

– the damage caused to neighboring constructions and other underground structures implies repairs, thereby leading to additional inconveniences for the public;

– the surface of the road is damaged, which reduces its life from 30 years to approximately 10 years. In the long run, the backfill used to fill up the trench after the work may sag; the road has ditches and bumps which affects its appearance as well as the comfort and safety of users. In addition, if an alternative route is provided, it is not always suitable and the road is damaged by excessive traffic. In all these cases, it is necessary to undertake additional repair expenses;

– in winter or in bad weather conditions, the work needs to be stopped. Over a long period of time, the loss of time, money and prolonged inconveniences can be significant.

In addition, the macro-economic impacts must also be listed:

– the use of quarries to extract material (grit, for example) which is buried some kilometers inside leads to its scarcity and to the rise in price of products that depend on it (concrete, for example);

– dumping sites, which are getting costlier and rarer, could instead be used for burying material other than rubble. These constitute mainly natural non-polluted areas;

– as it is technically avoidable, transporting thousands of tons of earth (4 to 10 tons per linear meter of pipeline) over a distance of thousands of kilometers contributes to the wearing of roads; for example, a truck weighing 38 tons causes 400000 times more wear to a motorway than an ordinary car (Lenglet, 1992). In addition, the national energy budget increases.

Table 8.1 sums up the different nuisances mentioned above and states some parties that are affected by the inconvenience: local communities, residents and small businessmen (Aït-Aïssa, 1997). The model records a total of 16 nuisances for the residents of which 16 are at the town level and 11 for small businessmen. Some common concerns such as the nuisances related to the safety of people, traffic or noise and vibrations are distinguished. From an economic point of view, it is interesting to specify the nuisances, which can be monetarily quantified with the market rate, and the non-quantifiable nuisances, which may be evaluated by other methods (for example, analogy with comparable situations, etc.).

8.1.3. *Reduction in nuisance by trenchless techniques*

The main advantages of trenchless techniques compared to open trench methods are the elimination of the majority of nuisances mentioned earlier. These are presented in a schematic format in Table 8.2 as proposed by Berosch (1994).

Table 8.2 distinguishes between three main groups of advantages (for the site, the traffic and the residents) and indicates their effects on various parameters (cost, planning, execution, acceptance and environment). It is observed that the advantages of these techniques reduce, above all, the total cost, facilitate the planning of work and improve their acceptance. In this manner, the table can be used to measure the impact of one element of the list in the framework of an analysis of direct and social costs of a spread or a rehabilitation site.

Local communities	Residents	Small Businessmen	Nuisances	Quantifiable
			a) Traffic disruption	
	X	X	- site equipment traffic	
X	X		- diversions and delays	X
X	X	X	- vehicles, pedestrians' access	X
	X	X	- decrease in parking space	X
	X		- removal of bus and taxi stands	X
			b) Damage to the environment	
X	X	X	- noise and vibrations	
X	X		- air pollution	
	X		- dust and dirt	
X	X		- damage to the urban landscape	
X	X		- damage to the green areas	X
X		X	- storage of materials	X
			c) Risk of accidents	
X		X	- safety of workers	
X	X	X	- public safety	
X	X		- risk of car accidents	
X	X	X	- safety of children and elderly people	
			d) Economic impacts	
X	X		- decline in commercial activities	X
X	X	X	- damage caused to neighboring constructions	X
X		X	- deterioration of existing networks	X
X	X		- deterioration of the road	X
X		X	- temporary stoppage of work	

Table 8.1. *List of problems caused by urban underground sites*

Advantages	Reduction of the total cost	Planning for simplifying work	Execution of the structure made easy	Better acceptance ("brand image")	Protection of the environment
a) For the site					
- reduced volume of rubble (minimum use of dumps)	X	X	X	X	X
- very little backfill (minimum exploitation of quarries)	X	X	X	X	X
- very few equipment items at the site	X	X	X	X	X
- minimum accident hazards	X			X	
- fewer personnel	X	X	X		
- meteorological independence	X	X	X	X	
- fewer conflicts with other distributors	X	X	X	X	
b) For the traffic					
- lesser site traffic		X	X	X	X
- in general, without blocking the way (fewer time delays, no diversions required)	XO	X	X	X	
- better accessibility for pedestrians	O			X	
- parking not very inconvenient	O	X		X	
- minimum inconvenience for public transport	O	X		X	
- longer life span of the road	X			X	
c) For the residents					
- Less noise (limited to the working shaft)	O	X		X	X
- fewer or no vibrations	O	X	X	X	X
- fewer smells (exhaust gases, sewers)	O			X	X
- less dirt, dust	O	X		X	X
- appearance of areas maintained (pedestrian precinct, monuments)	XO			X	X
- protection of trees, gardens, parks	XO			X	X
- reduced loss of time	XO	X		X	
- less surface occupied	XO	X	X	X	

Table 8.2. *Advantages of trenchless techniques compared to open trenches (O: social cost, X: direct cost)*

8.1.4. *Methods for evaluating the social cost*

Today, the concept of social cost is raised more and more frequently. Its consideration remains ambiguous and approximate in view of the difficulties faced in comparing the social costs and direct costs using the same criterion.

In fact, the significance of non-quantifiable nuisances compiled earlier indicates a complex situation. The non-quantifiable nuisances are not sanctioned by the market and do not form part of any pricing system; thus, they cannot be included in the value of a project even though it is possible to estimate approximate price indexes in some cases.

In the following section, we will first present the methods used in contexts other than urban sites and then the adaptation of these methods to urban sites.

8.1.4.1. *Methods used in a context other than that of urban sites*

8.1.4.1.1. A recent problem in research

In the 1970s and 1980s, the concept of social cost was extensively used in the *economy of transport*, particularly to prove the economic efficiency of transport in urban communities. Calculations related to losses incurred by time delays and congestions as well as noise and disamenity costs were developed (Beauvais, 1978). The concept of consumption of time and space measured in $m^2 \times$ time was studied to lead to a space consumption cost (Marchand, 1984).

The question of monetary evaluation of the impacts of transportation on the *environment*, particularly of the integration of social costs on the economic calculus, has also recently been the subject of much research and practical applications.

In the context of high demand on the political plane, the INRETS – National institute of transportation and safety research – and CERTU – Centre for network, transportation, town planning and construction studies – have undertaken researches aimed at finding concrete solutions for the evaluation of social costs.

There are two methods frequently used to evaluate the social costs of transport: a mathematical approach known as "hedonistic pricing method" and a method organized in the form of surveys called "contingent evaluation".

These two approaches rely on the concept of "*consent to pay*", like most other available methods of monetary evaluation; the consent to pay is the sum that the individuals are ready to pay to avoid any nuisances or the sum that they are ready to accept as compensation.

8.1.4.1.2. Mathematical approach

The method of hedonistic pricing is often used to evaluate the impact of nuisances on the real estate market. It considers the building value of a real property "h" as dependant on all the benefits it receives.

These considerations have helped implement a mathematical calculation formula:

$$P_h = f(L_h, V_h, A_h, E_h)$$

where:

– “P_h” represents the value of the real property “h”,

– “L_h” all the characteristics peculiar to accommodation (number of rooms, etc.),

– “V_h” a group of proximity variables (density and social structure of population, etc.),

– “A_h” the parameters describing the accessibility conditions (proximity of shopping centers, schools, existence of public transport facilities, etc.),

– “E_h” the characteristics of the physical environment (pollution, noise levels, etc.).

8.1.4.1.3. Quota evaluation

By investigative procedures, the contingent evaluation method consists in asking the individuals the price they are ready to pay to enjoy the benefits or avoid a loss, or the compensatory amount they require to bear the inconveniences or give up the benefits. The contingent evaluation method is based on an approach comprising three major stages:

– detailed description of property to be evaluated and trade issues,

– obtaining the consent to pay or the willingness to accept of the person being surveyed through a simple questionnaire,

– analyzing the results so as to obtain consent to pay or the representative willingness to accept.

In order to minimize all forms of bias, this operation requires certain meticulousness during its implementation (simple and complete questionnaire, choice of sample, etc.).

8.1.4.1.4. Comments on the methods

Experience reveals some difficulties as well as some confusion in the use of methods to calculate social costs. The concept of consent to pay is the main cause. It creates numerous debates on monetization of external effects. In fact, for some, the social cost must reflect the property damage, which can be objectively measured through expenses that they induce. For others, this value must express a form of evaluation that a person assigns to a given effect, which naturally refers to relative measurements.

These monetary evaluation techniques also raise many methodological and practical problems. They tend to favor the simplest social groups, the public does not accept them totally and they require the collection of a large amount of data, etc.

8.1.4.2. *Approaches as part of urban underground sites*

For urban technical sites, the concept of social cost was the subject of numerous studies in some countries (Germany, England, the USA, France) so as to be integrated into the aggregate value of a project.

8.1.4.2.1. The German method

During a workshop session at the "NO DIG 97" social costs conference at Genoa, Bielecki and Schreyer presented a summary of their work relating to the development of a method for the implementation of a pipeline project.

The principle of the German approach is based on the control of a primary need, i.e. "the choice of the method and technique best suited to the environment when we decide to launch the installation or renovation work of the pipeline".

Some rules are proposed with the aim to be able to make comparisons between the open trench techniques and the trenchless techniques. The choice of the technique is made with the help of certain criteria (economic, legal, environmental and technical). The criteria are listed in a simple and comprehensible checklist and can be validated in two ways:

a) in the form of indices,

b) in monetary form.

Thus, users of the charter must be able to choose the proper technique, the proper contractor and a planning method suited to the requirements of the site in the surrounding environment.

8.1.4.2.2. English methods

In England, Vickridge proposes the setting up of a national legislature to monitor the social costs ("NO DIG 97" conference at Genoa). The steps proposed for the monitoring of social costs encourage the use of trenchless technology. In fact, these techniques help considerably reduce the negative effects on the environment and in this case the social costs.

Vickridge has distinguished between four priorities:

a) the use of public funds to reduce environmental adversities,

b) the standardization of the concept of equalizing claims to all those affected by nuisances caused by the site,

c) the adjustment of working time on the roads,

d) the standardization of taxation on the environment at urban sites.

At the NO DIG 97 conference, Robinson and Thomson presented another approach. The “digital and analytical methods” were developed with the aim of integrating the social costs in the comprehensive tax allocation scheme (excluding the total cost of work). This method can help make a reliable estimate of social costs using a procedure capable of correctly comparing calculations.

The sensitivity of the approach depends on the reliability of the type of data collected. It must be complete and precise.

Furthermore, it is necessary to provide the calculation coefficients to be able to validate the model. Actual costs can be used for quantifiable data; other additional data benefit from the coefficients developed during the discussions with the public community involved in the project (residents, municipality, companies, etc.).

Data and coefficients allow the use of a network of matrices to develop and analyze various scenarios. The scenario with the lowest score is selected; it reduces the inconveniences created by the site as well as the social costs.

The Jason model has been was applied successfully to several major urban underground technical installations.

8.1.4.2.3. American context

During the NO DIG 97 conference, Sterling presented a paper dealing with the indirect costs of an installation and repair project of underground pipelines in urban areas. The problems propose to assess the value of the public land before and after work, when a decision needs to be taken for the construction of a site. To do so, the studies are carried out based on three main themes:

a) minimizing social costs incurred by work and road traffic congestion,
b) management of underground public land,
c) management of the life cycle and pavements in terms of total cost.

8.1.4.2.4. French method

In France, no global method exists to combat nuisances generated by urban sites. The open trenches are commonplace in towns and cause enormous management problems and the progress of the work is confronted with obstacles that cannot be ignored (response of residents, petition against the failure to respect deadlines, etc.) due to negative effects on the environment.

With the experience of the transport economy, Angot (1991) has proposed an assessment of some social costs caused by congestion:

a) the total cost of congestion based on the hypothesis of a road remaining completely closed, thus leading to the use of a diversion,

b) the cost of delays due to a lack of traffic diversions,
c) additional cost of fuel caused by congestion,
d) use of parking space by those working on the site.

For other social costs like the social cost of noise, pollution, site hazards and loss in commercial activities, it follows the example of social costs defined abroad, particularly in Germany. Unfortunately, no practical exercise was undertaken on the ground to test and validate these propositions.

8.1.4.3. *Comparison methodology for the costs of trench and trenchless techniques*

In the absence of an official method for evaluating the social costs, Aît-Aïssa (1997) proposed a decision support method by comparing the cost of work with the trench and trenchless techniques.

Control over the inconveniences caused by urban sites is a well-known necessity in today's world. The method suggested by Aît-Aïssa consists of helping decision-makers choose operators using techniques that reduce considerably the nuisances and inconveniences of installation and repair sites for urban underground pipelines. The tools are easy to use and facilitate the task of technical services to evaluate the social cost of a site. All that is needed to be done is to fill up the multicriteria forms presented on the next three pages, which contain statistical data to compare different situations and conditions of the site.

Firstly, the general data of the site must be supplied to better comprehend the context of the project: characteristics of the road network (Table 8.3), type of work to be carried out (Table 8.4), duration of work (Table 8.5), density of the urban underground (Table 8.6), town boundaries considered (Table 8.7) and existing urban activities (Table 8.8).

<table>
<tr><th colspan="6">1. Characteristics of the road</th></tr>
<tr><td>Name of the road</td><td colspan="5"></td></tr>
<tr><td>Width</td><td></td><td colspan="2">Length</td><td colspan="2"></td></tr>
<tr><td>Pedestrianised</td><td></td><td>Motorised</td><td></td><td>Both</td><td></td></tr>
<tr><td colspan="3">Daily traffic</td><td>LD</td><td>D</td><td>VD</td></tr>
<tr><td colspan="3">Peak hours traffic</td><td>LD</td><td>D</td><td>VD</td></tr>
<tr><td>Main line</td><td></td><td colspan="2">Secondary road</td><td colspan="2"></td></tr>
</table>

LD=Less Dense D=Dense VD=Very Dense

Table 8.3. *Road characteristics*

<table>
<tr><th colspan="6">2. Type of work to be carried out</th></tr>
<tr><td colspan="2">Installation of pipes</td><td></td><td colspan="2">Renovation</td><td></td></tr>
<tr><td colspan="4">Number of linear meters</td><td colspan="2"></td></tr>
<tr><td colspan="4">Diameter of the pipelines</td><td colspan="2"></td></tr>
<tr><td>Drainage</td><td></td><td>Gas</td><td></td><td>Electricity</td><td></td></tr>
<tr><td>Telecommunications</td><td></td><td>Heating</td><td></td><td>Other</td><td></td></tr>
<tr><td colspan="2">Technical possibilities</td><td colspan="2">TET-Y/N</td><td colspan="2">TST-Y/N</td></tr>
</table>

TET=Trench Techniques Y=Yes
TST=Trenchless Techniques N=No

Table 8.4. *Type of work to be carried out*

3. Duration of work	
with open trench technique	
with trenchless technique	

Table 8.5. *Duration of the work*

<table>
<tr><th colspan="6">4. Urban underground density</th></tr>
<tr><td>Not very dense</td><td></td><td>Dense</td><td></td><td>Highly dense</td><td></td></tr>
<tr><td colspan="4">Types of existing networks</td><td colspan="2"></td></tr>
</table>

Table 8.6. *Urban underground density*

<table>
<tr><th colspan="6">5. Town boundaries considered</th></tr>
<tr><td>Not very dense</td><td></td><td>Dense</td><td></td><td>Highly dense</td><td></td></tr>
<tr><td colspan="3">Total area</td><td colspan="3"></td></tr>
</table>

Table 8.7. *Town boundaries considered*

6. Existing urban activities				
Business	0	1	2	3
Residences	0	1	2	3
Offices	0	1	2	3
Services	0	1	2	3
TOTAL = /12				

0 = Non-existent 1 = Low 2 = Average 3 = High Low urban activity if 3/12, Average if 6/12 and high if 12/12

Table 8.8. *Existing urban activities*

Next, the degree of inconveniences are identified in the following three situations to be compared: the initial state before work (Table 8.9), work using trench techniques (Table 8.10) and work using trenchless techniques (Table 8.11). A cross is marked in the table against the "less sensitive", "sensitive" or "very sensitive" categories of various inconveniences considered. They are classified into four categories:

- traffic disruption,
- damage to the environment,
- risk of accidents,
- economic impacts.

NOTE: these tables are easily comparable. They are reproduced here in their entirety because it is convenient to fill in all three so as to have the three grids at our disposal for each of the three situations (initial state, trench and trenchless work), which can help to compile the following summary tables.

7.a. DEGREE OF INCONVENIENCE (Initial condition before work)	Not very sensitive	Sensitive	Very sensitive				
Inconvenience				Inconvenience			
Traffic disruption							
Disruption to vehicles				Vehicles' access			
Deviations and delays				Pedestrian's access			
Damage to the environment							
Noise and vibrations				Air pollution			
Dust and dirt				Deterioration to the urban landscape			
Damage to the green areas							
Risk of accidents							
Risk of car accidents				Public safety			
Safety of children and elderly people							
Economic impacts							
Decline in commercial activities				Damage caused to constructions			
Deterioration of existing networks				Deterioration of the road			

Table 8.9. *Degree of inconvenience – initial condition before work*

7.b. DEGREE OF INCONVENIENCE (for work in trenches TET)	Not very sensitive	Sensitive	Very sensitive				
Inconvenience				Inconvenience			
Traffic disruption							
Disruption to vehicles				Site equipment traffic			
Diversions and delays				Vehicles' access			
Pedestrians' access				Decrease in parking space			
Reduction of rights of way				Removal of bus and taxi stands			
Damage to the environment							
Noise and vibrations				Air pollution			
Dust and dirt				Deterioration to the urban landscape			
Damage to the green areas				Storage of materials			
Risk of accidents							
Safety of workers				Public safety			
Risk of car accidents				Safety of children and elderly people			
Economic impacts							
Decline in commercial activities				Damage caused to constructions			
Deterioration of existing networks				Deterioration of the road			
Duration of the work				Temporary stoppage of work			

Table 8.10. *Degree of inconvenience – for work in trenches*

7.c. DEGREE OF INCONVENIENCE (for work in trenches TST)	Not very sensitive	Sensitive	Very sensitive				
Inconvenience	Not very sensitive	Sensitive	Very sensitive	Inconvenience	Not very sensitive	Sensitive	Very sensitive
Traffic disruption							
Disruption to vehicles				Site equipment traffic			
Diversions and delays				Vehicles' access			
Pedestrians' access				Decrease in parking space			
Reduction of rights of way				Removal of bus and taxi stands			
Damage to the environment							
Noise and vibrations				Air pollution			
Dust and dirt				Deterioration to the urban landscape			
Damage to the green areas				Storage of materials			
Risk of accidents							
Safety of workers				Public safety			
Risk of car accidents				Safety of children and elderly people			
Economic impacts							
Decline in commercial activities				Damage caused to constructions			
Deterioration of existing networks				Deterioration of the road			
Duration of the work				Temporary stoppage of work			

Table 8.11. *Degree of inconvenience – for trenchless work*

Finally, several summary tables help compare the costs of two techniques.

Table 8.12 summarizes the results of previous tables in the form of the percentage of different degrees of inconvenience in the three situations "initial state, trench and trenchless work"; it must be filled in by determining, using Tables 8.9 to 8.11, the proportion (in %) of categories classified as:

– "not very sensitive": 16 categories,

– "sensitive": 24 categories,

– "very sensitive": 24 categories.

Table 8.12 defines the most appropriate "inconvenience category" for the situation, which means the one that corresponds to the highest percentage of each of the categories represented.

Percentage of inconvenience (*)			
Initial state	%	%	%
If TST	%	%	%
If TET	%	%	%

Less sensitive Sensitive Very sensitive

(*) Summary
of tables 8.9, 8.10 et 8.11

TST: Trenchless Techniques
TET: Trench Techniques

Table 8.12. *Percentage of inconvenience*

Table 8.13 presents the coefficients α_{CS} to be applied to the Direct cost CD to calculate the social cost (see Appendix 1, glossary of symbols):

$$CS = \alpha_{CS}.CD$$

Coefficients obtained from some experiments are proposed by default.

9. Multiplying coefficients α_{cs} (en %)			
CD: Direct Costs			
0 < CD < 0.15 M€	2%	6%	10%
0.15 M€ < CD < 0.6 M€	4%	9%	14%
0.6 M€ < CD < 1.2 M€	6%	11%	17%
1.2 M€ < CD < 1.8 M€	8%	13%	20%
1.8 M€ < CD < 2.4 M€	9%	15%	22%
2.4 M€ < CD < 3 M€	10%	16%	24%

Table 8.13. *Multiplying coefficient α_{CS} (in %)*

Determined from experimental results, the multiplying coefficients α_{CS} are proposed "by default"; they will be applied to the most suitable "inconvenience category" (obtained from Table 8.12). For example, if the trench technique records the highest percentage in the "very sensitive" cell and the cost of work lies between 0 and 0.15 M€, we will obtain:

– the social cost CS by multiplying the direct cost of work DC linked to this technique by 0.10: $CS = \alpha_{CS}.CD$, .CD,

– the total cost TC, by multiplying the direct cost of work by 0.10: CD = (1 + α_{CS}).CD

Tables 8.14 and 8.15 compare the results obtained for the two techniques and show which technique is most suitable for the project.

10. Cost of indexed operations		
	Trenchless Technique	Trench Technique
CD Direct Costs		
Coefficients α_{CS}		
Total Cost = $(1+\alpha_{CS})$.CD		

Table 8.14. *Cost of the indexed operation*

NOTE: the use of "default" coefficients is proposed so as to indirectly and approximately measure the total worth of the project. It is thus advised to study the results of the "Total Cost" column with a lot of flexibility.

11. Assessment and operational choice
Are the Trenchless Techniques more expensive? YES □ NO □
Are the results of the characteristics of the environment (density of the underground soil, urban activity, sensitivity of the environment to inconvenience, etc.) decisive factors in the use of Trenchless Techniques? YES □ NO □

Table 8.15. *Assessment and operational choice*

Thus, by combining several technical, environmental and urbanistic components, the user must be able to make choices for the least inconvenient solutions suitable to the local situations and get an initial estimate of the total cost of work by assigning a "social cost" coefficient to the direct cost of the operation.

8.1.5. *Other suggestions to reduce the social cost*

It is possible to envisage intermediary solutions following the example of the English approach, which proposes the setting up of additional measures to better manage the social costs and encourage the use of the trenchless technique.

At this stage of our studies, we can propose – *at the scale of each community* – the implementation of guidelines favoring the reduction of inconvenience created by urban sites, such as:

a) preparing susceptibility maps,

b) financial incentives

8.1.5.1. *Susceptibility maps*

Preparing a "susceptibility map for nuisances and inconvenience" is recommended in cities and large towns. This map will help forewarn the intervener of the susceptibility of the site and the police authorities could impose certain execution methods in areas with "high susceptibility" or in some cases stop work, which is not directly connected to the services of nearby residents. It will be used in a systematic way in the initial stages of all operations carried out in the public domain. The planning authority will be responsible for the preparation of susceptibility maps.

8.1.5.2. *Financial incentives*

8.1.5.2.1. Environmental taxation

The use of public roads for the installation or repair of urban pipelines will be subject to payments for the entire duration of the work. The rates will vary according to the occupied area and the duration of operations. The rates may also vary according to the susceptibility of the site as defined by the previous mapping.

For the time being, the current regulations providing for the possibility of a royalty payment could be used in order to obtain a road maintenance permit. In metropolitan cities, the use of the local area is often subject to payments, for example for parking along the road where the amount is variable depending on the time and place of parking.

It seems probable that urban sites will be handled in the same way as already done in Japan by the *Road Occupancy Charge* (Fujita, 1990) and recommended in the United Kingdom with the creation of the *Road space rental tax*.

8.1.5.2.2. Tax relief

Tax relief may be given to the operators who demonstrate the use of techniques that help significantly reduce the problems and inconveniences caused by the work.

8.1.6. *Conclusions*

The analysis presented in this chapter shows that the social benefit in using trenchless techniques is considerable.

In fact, the public actively opposes problems created by an open trench in an urban area. The investigations undertaken with the public and the road workers have revealed that people want to eliminate the problems as well as their difficulty in anticipating their cost.

Various methods of evaluating the social cost have been summarized in Table 8.16, but the hypotheses considered in the models indicate the limitations of these approaches, which are still at the experimental stage.

In practice, in the absence of an evaluation method for predominantly known social costs and the near failure to take into account the problems during work at urban sites, the objective of the approach presented here is to propose a simplified method in order to facilitate a better understanding of the feasibility of the work with trenchless techniques.

Method Country	**Mathematical formulation**	**Comparison method**	**Others**
Germany		Comparison between the installation of pipelines using the open trench and trenchless techniques (economic, environmental and technical aspects) (Bielecki and Schreyer, 1997)	
United Kingdom	Digital model for evaluating the social cost for the calculation of the total cost (Robinson and Thompson, 1997)		Proposal for setting up of a national legislation to monitor the social costs (Vickridge, 1997)
United States	Implementation of a method to fully evaluate the social cost of an urban site (Sterling, 1997)		
France	Determination of the social cost uniquely linked to the traffic (Angot, 1991)	Comparison of the working cost with the trench and trenchless techniques (Ait-Aïssa, 1997)	

Table 8.16. *Summary of evaluation methods of social costs*

8.2. Contractual aspects: objectives and success factors

This section, concerning the contractual aspects of microtunneling markets, includes a certain number of recommendations intended to help in:

a) correctly contractualising a microtunneling project depending on its context (see paragraph 8.2.1), bearing in mind an optimal realization of content/quality/environment/cost/timeframe of the project,

b) preparing a tender document (DCE) and an appropriate consultation regulation (RC) (see paragraph 8.2.2), to help the contractor contractually commit and express his professionalism,

c) submit compliant and pertinent offers (and to the requirements of applications) so as to meet the expectations of the owner (see paragraph 8.2.3).

NOTE: in this discussion, all processes of construction, renovation or replacement by trenchless techniques of an underground structure using a shaft (chain of pipes or cylinders) inserted through an entry point (generally an attack shaft) up to an exit point (recovery shaft) are known as microtunneling. By extension, the current document also applies to the pseudo-accessible and accessible fields: microtunneling work (∅ 1200 to 1800 mm) and other larger sites in some cases.

8.2.1. *Proper contractualisation of a microtunneling project*

A microtunneling project will be properly studied, contratualised and then carried out if:

– the roles of the owner, engineer and contractor are well defined (see paragraph 8.2.1.1),

– the risks of the project are well managed (see paragraph 8.2.1.2),

– there is sufficient knowledge of the nature and use of the ground surrounding the microtunnels (see paragraph 8.2.1.3),

– allotment of work and contracting are suitable (see paragraph 8.2.1.4).

8.2.1.1. *Well defined respective roles*

The owner:

– *analyzes* his own requirements and expectations and *formalizes* his program constraints (particularly operational ones) and the service expected (quality of use) for the microtunnel;

– *finances* all the studies that are considered a genuine investment (particularly geotechnical investigations);

– *approves* the reference project established by the engineer, particularly the identification of risks that can occur at this stage; then, an approval, which is decided through the DCE, will transfer all the technical, financial and legal responsibilities to the contractor;

– *selects* and contractualises, with the contractor, the proposal of the engineer.

Note: the owner may encourage competitors to submit detailed offers to compensate for offers that are not retained at their true value;

– *finances* the major modifications, if any, in the program, which exceed the contractual framework finalized between the parties, particularly in relation to geotechnical hazards;

– *recovers* the ownership of the commissioned structure after the acceptance of work and the administrative and financial payment of the market(s) concerned.

The engineer:

– *collects and prioritizes* all the technical, physical, administrative and environmental constraints acting on the microtunnel;

– *carries out and/or helps carry out* with the help of the appropriate subcontractors advanced studies, particularly on the geotechnical plane, after having *analyzed* and *selected* the offers related to the qualifications of geological and geotechnical study material;

– *sizes* the material beforehand, *recommends* the procedures, *identifies and evaluates* the inherent risks of the project as part of the "reference project".
Note: the reference project is intended to be detailed, completed, or even partially reworked by the contractor (in all cases he must acquire it within the framework of his offer);

– *provides* for, using a schedule of due dates and a suitable planning (scheduling), sufficient time delays for the entire operation and prepares the contractualisation from the reference project, which serves as basis for the creation of the tender document (possible applications, DCE).
Note: this reference project helps the contractor to contractually get involved by integrating and supplementing the identified risks;

– *analyzes* the offers, particularly those related to the qualifications of competitors in terms of products and procedures and, if need be, geotechnical studies;

– *adjusts* the contract proposals when required if the reference project was completed since its approval by the owner (additional optimization, competitor variants, etc.);

– *implements* the requirements of the market (required quality) during the construction and acceptance phase of the microtunnel.

The contractor:

– *acquires* the studies of the engineer, updates the risk analysis by clarifying his own risk taking, *completes* the reference project, if need be;

– *takes responsibility* for technical contents, the delays and cost with products and procedures established on the basis of the reference project.

Note: the contractor must not hesitate to partially reorganize the reference project if the risks and/or the respect for costs and deadlines require it;

– *identifies* and *undertakes*, if need be (on the basis of a specific price or on his own initiative) further studies, particularly geotechnical ones.

– *carries* out the work envisaged in the contract as a professional (quality obtained, in accordance with the quality required and contractually defined by the engineer and consistent with the quality expected by the owner).

Figure 8.1 shows the sequence of activities and the role of each representative described above required to complete a microtunnel complying with the expectations of the owner.

Here are some remarks on Figure 8.1: "the casing" on the "reference project" display shows the degree of more or less significant improvements in the reference project, which could be according to the following cases:

– either not very detailed (general concept level), for example in the case of competitions, and the competitors followed by the contractors will bring in a lot of grey matter;

– or quite detailed (detailed level) in the case of standard invitations to tender, with several occasions and possibilities of optimization and variants nevertheless;

– or very detailed (detailed design level) in the case of an order during the work or negotiated contracts with extremely short deadlines, the solution being practically stopped by the engineer.

8.2.1.2. *Appropriate risk management*

8.2.1.2.1. Risks to be considered by the owner

The following are examples of risks to be considered *initially* by the owner and corresponding to major potential modifications of the design program:

– scheduling of unsuitable operations;

– studies and geotechnical and hydrogeological drilling being totally reassessed by unexpected discoveries in the ground (archaeological vestiges, old mines and garbage dumps, perched or artesian water tables, radioactive materials, etc.);

– inappropriate consultation procedure, ineffective play of the competitors, imprecise or incomplete offers, fruitless invitation to tenders, insurance covers unsuitable for the guarantees, bad choice of contractors or sub-contractors, business failures (liquidation);

– other political, economic, health and safety or environmental risks linked to the change in the legislation, the development of administrative procedures (public purpose, water act, appeal, etc.), the interfaces with adjacent and/or simultaneous sites, dealer networks or building foundations being too close, specific complaints of residents, the concrete manifestation of these risks being able to interrupt, postpone or adjourn and thus make the operations more complex in all cases.

The owner must finance the consequences of such risks because they are caused by major modifications in the design program, can lead to additional working costs and also to direct or induced costs for contractorship.

8.2.1.2.2. Risks to be formalized by the engineer

This deals with technical, legal, administrative, financial and environmental risks identified in the reference project from studies undertaken at this stage. These risks must be included in the project estimation as well as in the estimation of the market(s) concerned after being developed according to their gravity and their probability of occurrence.

These risks must be clearly explained in the consultation documents particularly in the general fascicle of the terms and conditions and in the concerned wordings of the Unit Price Table (see paragraph 8.2.2.2). Their impact on the consultation procedure may be substantial:

– standard break up of the aggregate price or contract payment,

– well defined fixed rate (some sensible contract prices paid …),

– non-competitive bid contract (after qualification) with some companies who can control some major risks.

NOTE: the case of private contracts will have to be additionally specified.

8.2.1.2.3. Risks to be borne by the contractor

After appropriating the reference project and having completed or even partially reorganized it, the competitors should state in their technical submission (see paragraph 8.2.3.3) the technical, legal, administrative, financial and environmental risks that they have taken into account in their offer.

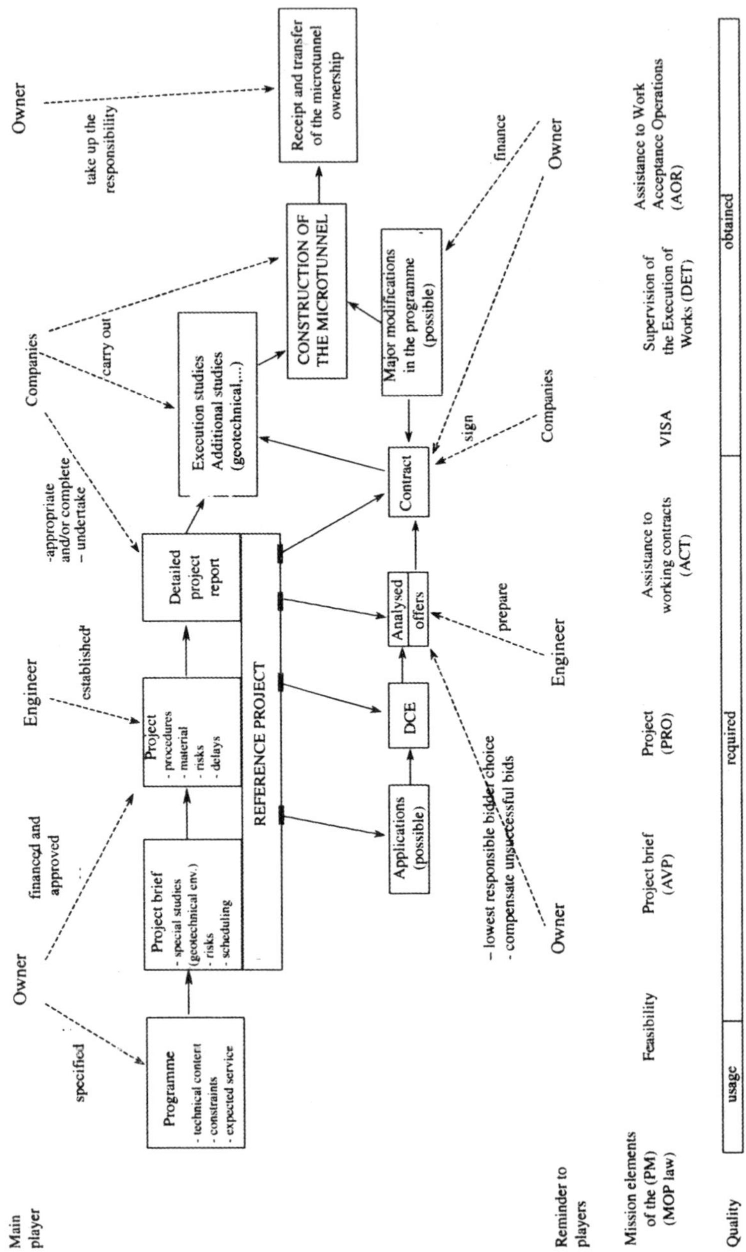

Figure 8.1. *Sequence of activities and the role of each party*

These risks must be included in the contract price after development according to their gravity and their probability of occurrence.

8.2.1.3. *Knowledge of the structure and underground use*

The success of a microtunnel project depends mainly on as precise a knowledge as possible of the nature of the ground in which the microtunneling works are being executed. Regarding the soil studies, the reader will have to refer to Chapter 4 to define the nature of the hydrogeological and geotechnical studies to be carried out:

– in particular, additional studies to be carried out by the engineer during the consultation phase if the project deadlines require it;

– additional hydrogeological and geotechnical studies to be carried out by the contractor during the execution studies.

8.2.1.4. *Suitable allotment and contracting*

The microtunneling deals were passed on to the contractor at the end of a consultation procedure (limited after qualification, as open tenders or negotiated procedures) defined by the owner according to its status and wordings applicable to it (member of the public subject or not to the Code of public contracts).

The works must form part of either an independent project contract, or a specific incidental lot within a larger market so as to:

a) enable the "microtunneling companies" to list realistic rates by consideration in relation to other members of the group within a framework of limited tendering,

b) to be able to prequalify the "microtunneling company" candidates before inviting competitive bids and before forming the groups within the framework of limited tendering,

c) to reassure the owner of the capacity of the contractor to meet the contractual requirements.

Nevertheless, it is recommended that the microtunnels require:

– deep shafts, particularly under high levels of unconfined groundwater,

– curve blasting or requiring intermediary stations *a priori*,

– prior treatment of the soil by injection or freezing,

– complex confinement modes (compressed air + earth pressure balance, confinement using bentonite slurry + mechanical masks, etc.),

– or if the share exceeds 30% of the total market value, it automatically becomes the object of an independent contract; this helps in specific consultations with the successful tenderers with appropriate expertise (see paragraph 8.2.3.1).

In other cases, subcontracting by the market tenderer to an appropriate microtunneling company is likely (specialized subsidiary, actually qualified local company, etc.).

8.2.2. *Establishment of appropriate tender documents and a consultation regulation*

A consultation document of microtunneling companies and its associated consultation regulations will be useful if:

– the consultation strategy is well defined and validated by the owner (see paragraph 8.2.2.1),

– every administrative or technical document includes the necessary specifications (see paragraph 8.2.2.2),

– the project is properly described (see paragraph 8.2.2.3),

– the products (pipes, joints, injection) are correctly sized and adapted (see paragraph 8.2.2.4),

– the microtunneling procedures are well controlled (see paragraph 8.2.2.5).

8.2.2.1. *Tender documents based on a defined strategy*

The compilation of the consultation document of companies is based on a preliminary reflection on the consultation strategy:

a) scheduling strategy (when to consult according to the progress of the project studies; see Figure 8.1),

b) strategy for risk sharing and management of litigations (see paragraph 8.2.1.2),

c) strategy for allotment, consultation on call for tenders (see paragraph 8.2.1.4).

To remain within the framework of the reference project, the structure and the wordings of the tender documents must be deliberately generic in relation to the material and procedures used. Only some particular specifications (in terms of objectives to be attained, functions to be provided and held according to the risks, etc.) must be detailed.

In fact, this deals with widening the choice of materials and procedures depending on the constraints specific to each insurmountable open-air obstacle: Instead of risking the inept stipulation of procedures, it is better that the contractor specifies the procedures according to his know-how and his own methods of execution. This helps assign a further sense of responsibility to the contractor and encourages him to innovate.

As a result, the tender documents specify, depending on the case:

– either the functions of a microtunnel to be provided,

– or the options retained by the engineer at the end of the project studies (materials and procedures).

However, in all cases, the tender documents are first obligatory for the competitors, due to the organizational chart of the quality assurance plan (see paragraph 8.2.2.2) and the technical submission attached to the offer (see paragraph 8.2.3.3), then for the contractor following his quality assurance plan:

– on the one hand, to develop his own method of working and monitoring,

– on the other hand, to submit for the acceptance of the engineer of the execution methods that he retains, as well as the assessments that he employs to ensure the required quality is achieved.

8.2.2.2. *Specifications adapted to every item of the tender documents*

Tables 8.17, 8.18 and 8.19 propose some guidelines for drafting specifications for each significant item linked to the consultation procedure.

8.2.2.3. *A properly described project*

In order to describe in the tender documents the crossing of insurmountable open-air obstacles, the engineer must:

– indicate the situation of one or more obstacles and the layout of one or more microtunnels,

– specify the fluids to be carried and/or dry systems, the available rights of way and the accessibility conditions for the entry and exit shafts,

– supply the characteristics of the microtunnel to be constructed and the functions that it must provide by summing up the data to be provided to the competitors through Table 8.20 (to be adapted to the context before being completed).

Item	Guidelines and recommendations for the drafting of specifications	Comments
Public notice of appeal to the Competitors (AAC)	Specify the qualifications (see paragraph 8.2.3.1) and references (see paragraph 8.2.3.2) required with equivalents for foreign candidates	- prohibit or authorize the variants - choose an appropriate allotment - determine the legal form of the owner with whom the deal will be concluded: contractor with authorized subcontractors, joint or interdependent groups
Consultation regulations (RC)	State clearly: - prioritized judgment criteria, - possibility of variants or otherwise, - list of items (contractual or otherwise), - technical submission.	Preference to the lowest responsible bidder and not to the lowest bidder
Deed of covenant (AE)	Specify the realistic deadlines	Deadlines including the major risks
Specific administrative clauses (CCAP)	- content of rates – work evaluation methods – Extension of deadlines in case of bad weather, - specify possible simultaneous work, - specify the constraints (public highways, networks ...), - specific guarantees/new materials and processes, - specific guarantee/performance and results, - insurance.	The CCAP is generally aimed at CCAG-Works as a general document applicable to the contract Dispensations to CCAG-Works, to the standards and the CCTG to be indicated in the last article of the CCAP

Table 8.17. *Main specifications for the administrative items of tender documents*

Item	Guidelines and recommendations for the drafting of specifications	Comments
Specific technical clauses (CCTP) "General fascicle"	General prescriptions in terms of: - organization and preparation of works, - implementation program, - constraints linked to the networks and road traffic, - protection of the environment, implementation studies, staking out …	
Specific technical clauses (CCTP) "Microtunnel fascicle"	- origin and specification of materials (see paragraph 8.2.2.4), - implementation and monitoring terms (see paragraph 8.2.2.5), - acceptance and completion.	It may be necessary in some cases that the geotechnical document be contractual so that the competitors can commit themselves better As part of a competitive consultation, it is preferable to attach a reference CCTP to the tender documents technically translating the reference project mentioned in paragraph 8.2.1
Schedule of unit prices (BPU) **Detailed estimate (DE) of cost or breakdown of total and contract prices (DPGF)**	The consultation of microtunneling may be standard or in design-implementation (ready-made), it is better to attach a reference BPU and DE to the DCE (possibly without quantities)	The contractor must complete these documents in his tender from his own project studies

Table 8.18. *Main specifications for the technical items of tender documents*

NOTE: it is recommended that the technical specifications relating to the microtunnel(s) be formalized in a specific fascicle of the CCTP (microtunnel fascicle), which consists of general specifications (general fascicle), particularly if the microtunnel works are coupled with other works within the same contract.

Item	Guidelines and recommendations for the drafting of specifications	Comments
Organizational chart of the quality assurance plan (SOPAQ)	This deals with a standard space that needs to be filled up by every bidder, by stating the organization and the list of procedures envisaged to ensure that the contract terms are met. A complete SOPAQ is contractual	The contractor establishes the PAQ during the preparation period after signing of the contract. It is a living document, which is not designed to be contractualised
Organizational chart of environmental assurance plan (SOPAE)	From the SOPAE framework provided in the tender documents, the bidders must formalize the organization considered to guarantee environmental safety, by considering particularly the main constraints of sensitivity and potential impacts listed in the "Environment Instructions" (valuation of rubble from mucking, urban environment, mixing of polluted soil, drilling fluids …)	The contractor establishes the PAE during the preparation period after signing of the contract. The PAE is a living document, but which may be contractualised if the owner so wishes, particularly as part of an ISO 14001 approach. Similarly, the environmental instructions may be contractual
General security and health safety co-ordination plan (PGCSPS)	Ask the bidders to specify in their tender the measures taken in relation to: - reinforcement/strutting of shafts - safety near the machines	If the microtunneling and other simultaneous works justify it, the owner may implement a PGCSPS. In all cases, the contractor will establish the PPSPS during the preparation period

Table 8.19. *Main specifications of other items of the tender documents*

No. of the microtunnel	Pk	Functions that the microtunnel must provide	Constraints linked to the microtunnel	Cursory characteristics	Related drawings
- - -	- - -	- Fluids to be transported - Mechanical strength - Low cover - Unstable ground	- Dry networks to be protected - Rolling loads - Unconfined groundwater - Watertightness - Geometrical tolerance	- Lengths - Rate of flow or gradient - Diameters	

Table 8.20. *Description proposal of the microtunnel to be constructed*

8.2.2.4. *Correctly sized and adapted products*

It is based on all the geotechnical studies attached to the tender documents that the contractor undertakes on the capacity of materials to meet the constraints acting on the microtunnel, either during the construction or the operating phase, even if it means proposing materials (casing, joint, grout, etc.) other than the reference materials mentioned by the engineer in the tender documents, and compatible with the other prescriptions of the tender documents.

As the CCTG (fascicle 70) is very brief on the subject, the tender documents specify the line of measures contained in Chapter 6 of this book.

The contractor must specify the different characteristics of the pipes and cylinders used according to the project specifications (particularly the nominal diameter, required durability, etc.) and of the "supplier" documentation of pipes envisaged, particularly:

– material (concrete reinforced with or without sheet metal, sandstone, PVR, steel, HDPE, etc.);

– series (thickness, mechanical resistance, stiffness classification, etc.);

– joints between components and assembly procedures (fitting, welding, etc.);

– injection devices (lubrication, blocking, etc.).

The contractor will submit a design note of the mechanical dimensions of the pipes to the engineer. The pipes are sized according to the soil and the excess load (either static or rolling) that it must bear as well as the longitudinal thrusts that it must support. Furthermore, the pipes are calculated according to the radial stresses in case of a misalignment of one or more elements, particularly after a prolonged stoppage at the site (for example, swelling clay). In particular, the joint of each pipe is designed to resist the wear and tear caused by the compression/decompression cycles during each pushing sequence.

8.2.2.5. *Well defined and controlled microtunneling procedures*

It is by relying on all the geotechnical studies attached to the tender documents that the contractor undertakes the most suitable procedure to construct the microtunnel, even if it means suggesting procedures different from the reference procedure mentioned by the engineer in the tender documents, provided that they remain compatible with the other specifications in the tender documents.

The different procedures for microtunneling can be stated according to the following methods:

– *digging*: wheel canopy equipped with equipment suitable to the soil, etc.

– *confinement*: mechanical confinement by earth pressure, sludge pressure, compressed air composite solutions (not required if there is a good cohesion or pre-treatment of the soil, etc.)

– *transportation of rubble*: hydraulic mucking by confinement sludge, endless screw, etc.

– *assembly*: fitting and/or arc welding, etc.

– *penetration*: pushing using hydraulic jacks, reboring of the pilot microtunnel by permanent pipes, etc.

– *guiding*: differential pressure on the last pushed component, adjustable cylinders at the head of the microtunnel, etc.

– *geometrical control*: automatic guidance of parameters, raised by pipe, detection chip of the canopy, etc.

For every sub-procedure and common method affected by the project, the engineer, when required and the contractor in all cases, must specify at least:

a) the procedure envisaged: sequence of tasks: who will do it and how? (see Chapter 5);

b) the parameters governing the construction of the microtunnel (see Chapter 3 on this);

c) the inherent risks in the method and cases of non-implementation;

d) the actions in case of a malfunction (see section 3.2 on this);

e) the respective advantages and disadvantages.

8.2.3. *Presentation of compliant and pertinent offers by the contractor*

The items presented by the contractor in his bid will correspond to the expectations of the owner if the engineer has correctly specified the requirements (quality required), but also whether the contractor:

– has the appropriate qualifications (see paragraph 8.2.3.1),

– has suitable and sufficient references (see paragraph 8.2.3.2),

– submits a specific and complete technical submission (see paragraph 8.2.3.3).

8.2.3.1. *Appropriate qualifications*

In accordance with the classification of the FNTP, the qualifications required for constructing the microtunnel as well as the development work (shafts, pre-treatment, etc.) must be specified in the consultation documents, particularly in the AAC, under the following two imperative conditions:

– it must be ensured, when the AAC is published in JOCE, that the qualifications supplied by the contractor result, for example, from the FNTP or an equivalent document (to avoid discrimination against foreign organizations), by asking for a specific note showing the equivalence of foreign qualifications with those sought. In particular, anything being transported by rail will have the directions of the SNCF applied to it;

– a brief definition of the contents of the qualification must be given to make it comprehensible to a non-national that the number or acronym of the FNTP qualification does not give the correct information.

The qualifications defined by the FNTP are as follows. They are likely to change, so it is necessary to ensure the validity of the information mentioned below by referring to the FNTP website (www.fntp.fr), for example.

8.2.3.1.1. For microtunneling

a) 1.51 and 5.423: horizontal drilling for laying cables or pipelines,

b) 1.55 and 5.704: jacking of pipelines by spilling,

c) 1.6: underground work in urban sites using the trenchless techniques,

d) 1.7: underground work in non-urban natural sites using the trenchless techniques,

e) 5.705: laying of pipelines by boring machines.

8.2.3.1.2. For constructing shafts and pretreatment

a) 1.B: substantial special foundations and specific important execution procedures,

b) 1.2: foundations and strengthening of the soil by intermediary structures, particularly the 1.23 specializations: concrete, wooden or steel sheet-piles, or 1.240: cut cribs,

c) 1.3: specific execution procedures linked to special foundation works, particularly the 1.33 specialization: submerged concrete and concrete made in water by the injection of a special mortar,

d) 1.4: investigation, strengthening and sealing of soil, particularly the 1.42 specialization: injections for strengthening or sealing of soil carried out in open air or underground.

8.2.3.1.3. For geotechnical studies, if any (see NF P 94-500 standard)

a) G1: geotechnical feasibility study,

b) G2: geotechnical project study.

8.2.3.2. *Adequate and adapted references*

To gain a better insight into a company's experience, it is advisable to ask for:

a) precise references in terms of microtunneling, by attaching the most important certificates of competence,

b) possible general references in terms of underground works,

c) the curriculum vitae of technical or hierarchical personnel.

References acquired in the field of microtunneling

It is recommended that the candidates be asked the number of microtunneling sites constructed by them as well as the main characteristics of the important microtunneling sites that have been constructed by each of them:

– client,

– location,

– diameter(s),

– length,

– cost,

– pipe material(s),

– procedures used, difficulties encountered and solutions implemented,

– alignment defects, compaction, watertightness, overabundant friction, large blocs, etc.

Curriculum vitae of the technical and hierarchical personnel

Particular attention must be paid to the supervisory skills by insisting on seeing proof of the experience of the machine operator.

8.2.3.3. *A complete and definite technical submission*

The technical submission, whose contents are mentioned in the consultation regulations, consists of precisely describing the procedures and products retained (see paragraph 8.2.2.5), the intervention limits, as well as the apprehensible risks at the time of the offer, while projecting itself during the construction and operation phase of the microtunnel. It will include at least:

– a projected planning of execution;

– a development plan of site facilities (building, linking the energy sources to the networks, congestion of public highways for each shaft, etc.);

– a maximum power budget required by the facilities and the necessary energy sources;

– drawings and sections of the working shafts and the boring machine;

– an estimate of daily progress rates predicted;

– a material report of the boring machine used;

– the products selected (pipe, type of joints, sealing and injection slurry, etc.) in the form of product sheets;

– the follow-up methods and monitoring of digging, confinement and penetration;

– the procedures followed in case of an incident (encountering specific hydrogeological and geotechnical conditions, unusual wear of tools, etc.).

The contractor must make mention in his tender the measures he intends to take relating to:

– additional geotechnical and hydrogeological studies required in the concerned fields;

– nature of materials and characteristics of the shafts (pipes and/or cylinders), as well as their method of sizing;

– technical procedure and materials associated for constructing every microtunnel (particularly the digging method and the guidance system);

– "soil-digging" and "soil-casing" interactions (choice of confinement, frictional coefficients to be considered, etc.);

– construction procedures of working shafts, as well as the restoration of the shafts or construction of manholes in the microtunnel;

– validation of the procedures and materials after carrying out the geotechnical and hydrogeological studies deemed necessary;

– the nature, planning and frequency of inspections (inspecting the products when received, inspection during production, compliance tests, etc.);

– procedures to ensure the respect of geometric tolerances in the case of incidents at the site, hitches related to the nature of the soil (presence of large blocks, foundations or existing cables, unusual wear of tools, etc.), unexpected presence of unconfined groundwater;

– steps the contractor intends to take if differential compressions or unacceptable upheaval of the ground occurs;

– maintenance instructions for the microtunnel during operation.

PART II

Horizontal Drilling

Chapter 9

Introduction to Guidelines: Purpose and Fields of Application

9.1. General introduction of "the trenchless technology"

These recommendations apply to the construction of structures by horizontal drilling, which forms part of the "trenchless" techniques. These techniques are currently used in urban areas, in an era where environmental concerns play an ever-increasing role. In fact, this deals with creating new utilities or renovating existing water, sanitary drainage, electricity, gas or other networks by minimizing the impact of surface sites, thereby reducing the inconveniences caused to users in comparison with the work carried out using "trenching techniques", which means digging an open excavation along the full length of the area worked on.

Even if this usually deals with works that do not have the same spectacular character as large subway or urban motorway sites for example, their importance in terms of linear structures entirely justifies the great interest we take, in terms of economic consequences as well as their close intermingling with the social life.

We ought to first state a definition, which helps determine better the field of application of these works. Obviously, the term "trenchless" is the opposite of "trench" works, but it is additionally reserved for the installation of networks of small diameter, which we call "non man entry", i.e. a man cannot get into it in normal working conditions: we accept in general that the upper limit is about 1,200 mm diameter.

We will therefore be interested in underground structures whose construction requires the implementation of remote controlled techniques, as one cannot access them either from the surface ("trenchless"), or from the inside ("unvisitable").

It is common in the field of "trenchless digging" to distinguish between various procedures, for which the techniques used are very different and whose fields of application are equally diverse. Firstly, new construction projects and old renovation projects have to be distinguished.

a) The new structures involve the creation of networks where nothing exists and again for this, two categories can be considered corresponding to very different techniques:

– *microtunneling* is used for networks with diameters generally ranging from 300 to 2,500 mm and which can go up to 2,000 mm. The boring machines resemble Tunnel Boring Machines (TBMs) of large diameters, and have the special feature of being miniaturized and remote controlled, which means that they can be operated without any human intervention inside the machine. The machines operate along a linear trajectory at variable depths ranging from just a few meters to more than ten meters and along a length of approximately 100 to 150 m: thus they have to be installed through shafts dug from the surface up to the depth of the project. This enables the machines and its pipes to be sunk to the depth required for the project and then be recovered at the outlet.

– *horizontal drilling* is used in general for urban networks of small diameter (50 to 1,200 mm) as well as for pipelines of up to 1,000 mm in diameter. The technique is derived from traditional drilling with the added ability to locate the position of the drilling head in the plane and/or in depth and above all to correct the direction if there is a major deviation from the trajectory. It mostly relates to low depth networks (a few meters at the most) but can, in some cases with appropriate equipment, be used for installing pipes at greater depths. This is not covered in this discussion.

These two techniques resemble the horizontal methods. Better known than the horizontal methods is the non-horizontal method characterized by the installation of structures of shorter length (less than or equal to 50 meters) with discharge or excavation of soil and the entry and exit pits. They are economical and enable the installation of a permanent structure by inserting elements either pushed immediately or later if the drilling "holds".

Of the five "non-horizontal" methods, three use the lateral movement of the soil and two methods use the excavation of the soil.

The three non-horizontal drilling techniques by movement of the soil are:

– *the impact mole,*

– *the piling of closed tubes*

– *static jacking of guidance rods.*

The other two non-horizontal drilling methods using the excavation of the soil for installation are:

– *piling of open tubes,*

– *static jacking by augering.*

b) The renovation of old structures is used for existing networks whose ageing state means they no longer function properly. Thus, it is better to replace them either by creating new networks parallely (we are thus led to the two previous cases), either by trying to put them back into normal working condition according to their purpose (renovation), or finally by using the infrastructure to lay a new pipeline (replacement). Here too more than 25 different techniques can be distinguished, which are not discussed here.

9.2. History and characteristics of drilling methods

Impact moles were invented by the Polish in 1958 and revived by the Soviet Mines Institute in 1960. This rope-drilling spindle was then introduced at the end of the sixties in Western Europe, particularly in England and France as well as in the United States.

Figure 9.1. *Impact mole*

Its principle is very simple:

– after digging the entry and exit pits, a cylindrical frame equipped with a truncated conical head pulls a pipe or enables the laying of a pipe installed by pushing and pulling;

– after several passes, the mole creates a channel for placing the pipe of a maximum of 180 mm. According to the geological conditions, the distances recommended vary from 5 to 25 ml, and the rates from 5 to 20 ml/h.

The piling of closed tubes, a very old technique, employs the installation of "permanent" steel tubes closed at the end on the digging side, from the entry and exit pits.

The rates vary according to the soil, diameters, materials and personnel (from 1 to 50 ml/day).

The floor space requirement at the surface remains inside the pits from 8.50 to 10 ml taking into account the length of tubes as 6 ml and the length of the machine, and by the site installation of 30 to 50 m^2 required for storing pipes, handling tools, gantry crane, crane, truck, compressor and welding unit. The floor space requirement can be reduced by the use of very short tubes of less than 1.00 ml, but this is not cost-effective.

The static jacking technique by displacement of the soil during passing of "pilot" bars is being developed more and more nowadays. Also known as "push rod", the principle is also very simple.

Digging the shafts, a hydraulic unit jacks into the ground bars fixed one after the other, a 50 mm "pilot" drilling thus being created. These pilot bars, which have a very small diameter for sinking into the ground, are then equipped, at the end of the exit pit, with a bell whose return traction enlarges the hole drilled, thus enabling the pulling out of the pipe in steel, PVC or polyethylene.

These materials give rise to the horizontal drilling techniques mentioned later. These first three techniques use the moving of the soil for digging.

The piling of open tubes technique, also known as "tube driven by impact moles", uses the excavation of the soil (Figure 9.2). The "scrapping", a term used to designate the removal of cuttings, is ensured by hydrocuring or by compressed air or hydraulic thrusts.

The jacking is done by core sampling the ground using a pipe attached to a binder or a cutting lip. The continuous pipe is rammed using a pneumatic impact mole resembling the piling system of closed tubes.

Figure 9.2. *Driven tubes*

Furthermore, the field of application of the piling of open tubes is similar to closed tubes or spindles or crossing special points such as roads, highways, railways, canals, buildings and runways.

The diameters can vary from 200 mm to 3,000 mm.

Static jacking by manual borehole driver is also done by excavating the soil. In this case an endless screw discharges the mud return products. The screw is equipped with a rotating drilling head fixed around a shaft on which is wound a spiral welded iron sheet actually called auger. This technique has been available since the mid 19th century and production was first performed in the United States.

These plants help in the crossings of steel and concrete pipes with a PE and PVC casing. These plants also help create special connections.

The overall dimensions of the shaft vary depending on the line installation (L = 10 ml, l = 2.50 ml for a diameter of 300 mm and l = 4.50 ml for a diameter of 1,500 mm) or for connections (L = 2.40 ml and l = 1.40 ml). The operation of machines requires the installation of a hydraulic diesel power unit or a connection on a backhoe.

Some of these materials are currently dirigible and have a retractable digging head and a "down-the-hole hammer" ensuring an impact. This down-the-hole hammer is coupled to the rotation of the manual borehole driver for digging.

Horizontal drilling: in 1973, Martin Cherington combined the traditional horizontal drilling technique with the petroleum drilling technique known as "horizontal drilling". The first crossing of a river was completed in California. Since then, this technique is often known as *River Crossing Industry* in the United States. This method was introduced for the construction of pipelines requiring clearing of special obstacles such as rivers, mountains, canals, etc.

The field of application for horizontal drilling was then extended to installation in urban areas of various utilities such as the distribution of drinking water, electricity, gas and telecommunications. It also enables, for example, the installation of deep drains for the clean-up of waste dumps, soil investigation and in some cases for gravity wastewater systems.

Figure 9.3. *Drilling rig*

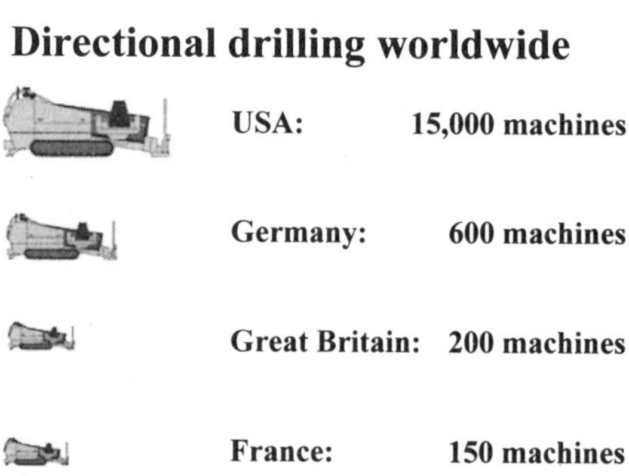

Figure 9.4. *Total number of drill rigs in the world*

Currently, horizontal drilling is a rapidly growing technique widely used in the United States where it was invented and in Germany. In France, after a slow start, horizontal drilling has seen a spectacular growth between 1990 and 2000.

Horizontal drilling, introduced in France in 1989, was used for the installation of flexible or rigid conduits with or without pits over variable lengths with powerful drill rigs and soil as already mentioned.

Figure 9.5. *Drilling rig*

The concept lies in the jacking of the pilot rod. A train of rods is inserted into the ground with the combined action of thrust and a rotating movement of a drilling head equipped with jetting water at the tip.

The technique can use compaction for non-sliding and not very hard soil and milling for rocky soil clumps. Nevertheless, for milling the dust needs to be collected.

In all cases, the objective is to construct a stable tunnel in the ground in which the work can take place.

Figure 9.6. *Pilot hole*

Figure 9.7. *Backreaming and pulling of the pipeline*

The installation of PE reels, welded rods or steel pipes made of welded rods is done by cable or pipe tractions when the pipe trains return after possible passage of many successive reborings.

The guidance is done by the flat end (angular head) located on the drilling head having a transmitter, with the receiver placed on the surface: the trajectory is straight due to the rotation of the pipes and a simultaneous thrust. It can be deflected only by using a thrust without rotation.

One of the advantages of this technique in comparison with the installation by open trenches is its speed of operation.

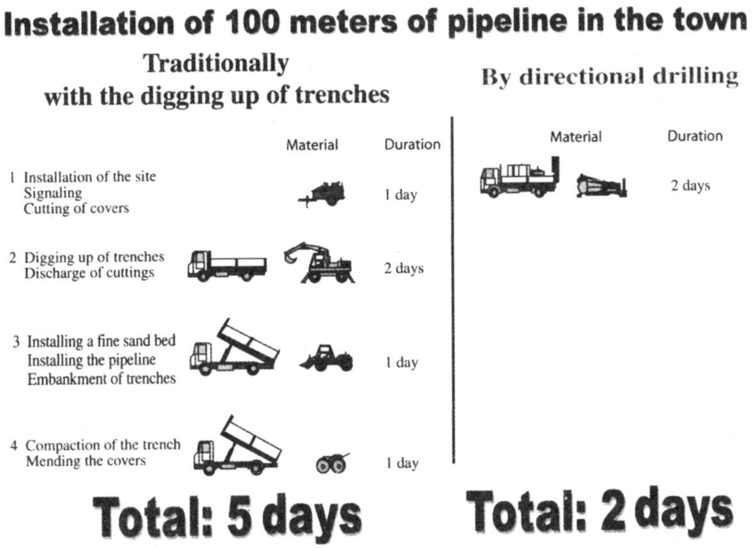

Figure 9.8. *Comparison between installation by trenches and by horizontal drilling*

The rate of progress of these equipment items is in the region of 10 to 300 ml/day for diameters of up to 1,200 mm.

In practice, 80% of the HDD drilling carried out is related to the installation of PE pipelines with a diameter not exceeding 200 mm. Table 9.1 summarizes the current chapter.

Classification	Special features	Group of techniques	Fields of application	Sections	Application lengths	Surface influence
Non-horizontal methods	Moving the soil	Spindle	Crossings and connections	From 30 to 180 mm	From 5 to 25 ml	Pits of 1 to 20 m^2 + compressor
		Piling of closed tubes	Crossing and connections	From 50 to 500 mm	Up to 50 ml	From 20 to 50 m^2
		Jacking of pilot rods	Crossings	From 30 to 250 mm	From 5 to 25 ml	2.40 ml * 1.50 ml Hydro-electric power station
	Hydraulic powerpack	Jacking of open tubes	Crossings	From 50 to 3,000 mm	Up to 70 ml	From 20 to 50 m^2
		Manual borehole driver jacking	Crossings and connections	From 50 to 1,500 mm	Up to 80 ml	≈ 25 m^2 in line and ≈ 5 m^2 across
Horizontal methods	Moving the soil	Dry horizontal drilling	Crossings and line installation	Less than 200 mm	Up to 240 ml	Drilling unit: 4 m^2, hydraulic powerpack: 2 m^2, compressor and possible pits of 3.30 m × 1.20 m
	By cutting and removal	Auger boring Microtunneler	Crossings and line installation	From 250 to 2,500 mm	Up to 150 ml	Pits varying from 2 m^2 cross section and 50 m^2 right-of-way to 20 m^2 and 150 m^2 right-of-way
		Slurry microtunneler	Crossings and line installation	From 100 to 1,200 mm	Up to 1,700 ml	Can be very significant
		Boring machine with hydraulic mucking	Crossings and line installations	From 250 to 2,500 mm	Up to 150 ml	Entry pits of 6 m^2 and exit pits of 3 m^2 and 150 m^2 right-of-way is much more significant

Table 9.1. *Characteristics of installation techniques for new structures*

9.3. Purpose of the recommendations and fields of application

The pipeline installation techniques diversified during the last decade with the arrival of the *Trenchless Technology* in France which groups together numerous solutions for the installation of all types of pipes and cables without digging trenches. These techniques relate to pipelines for water conveyance, sanitary drainage (waste water and rainwater), gas distribution and transportation networks, electric cables, telecommunication cables and fiber optic networks.

The techniques are numerous. They must be used wisely and taking into account the suitable solution in each case according to the soil into which the pipe and cylinder need to be laid (diameter and material), the precision required by the installation (gradient for sanitary drainage), the surface or underground congestion, etc.

These new and modern techniques have numerous advantages in comparison with trenching techniques. They considerably reduce the problems caused by work in urban sites (noise, mud, dust, traffic and parking restrictions, etc.) and its duration. They save on the movement of soil (cuttings, backfills) and thus on the natural materials. They sometimes help reduce the linear of the network because it is not necessary to get round obstacles or look – sometimes for considerable distances – for pre-existing passages.

They reduce the impact on surrounding structures, either buried or on the surface, on the roots of plants and street tree plantings, and avoid degrading the structure of roads. They reduce the risk of accidents caused by trenches for personnel as well as users of public property (pedestrians, bicycles, etc.).

These techniques were developed in different countries according to the requirements of the time, the desire to change habits and the capacity of the technicians to invent new solutions. It is generally the lack of space caused by urbanization which has led to the creation of this innovation. Elsewhere, it is the emergence of environmental issues and the desire to reduce inconvenience to residents and users of public property that has led to the use of these processes.

These “trenchless” techniques were progressively introduced in France at the end of the 1980s. It was in order to gain a better understanding of them and to develop and adapt them to the national context that the French committee for trenchless works – FSTT – part of the ISTT network (*International Society for Trenchless Technology*) was created. It currently groups together more than 25 national scientific associations set up over five continents. With the support of the Equipment ministry and its Research directorate (DRAST) and the Institute of applied research and experimentation in civil engineering, its workshops strive for this development, particularly through two national projects of research and development – the RERAU National Project for the renovation of networks, and the National Microtunnels Project, whose research concentrates on horizontal drilling.

So, it is generally down to the opinion leaders, project managers and prime contractors to promote and accelerate the use of these techniques. As for the additional costs compared to conventional trench techniques, one must obviously take into account the advantages that these procedures provide for the environment of the site and the quality of the finished work, as well as all the cases where these procedures help save considerably (if these techniques were globally more expensive, they would not have gained popularity and been implemented in some countries).

Derived from oilfield drilling, horizontal drilling is a technique that enables the installation of cables, ducts and pipelines through underground ways without digging trenches.

The crossing of special points is generally done by companies specializing in motorways, rivers, canals and railways where it often constitutes the best suitable solution, technically and economically.

In towns, the installation under road, rail and waterway networks by horizontal drilling is an alternative to traditional techniques, which provide considerable advantages for the environment:

– reducing inconvenience for residents and traffic (no trenches, no vehicles, no cuttings to be discharged, no damage to the roadway, etc.);

– significant time savings.

Even though horizontal drilling has inherited some specific tools and has been tested in various types of vertical drilling, it remains a recent technology still in development, and every year equipment manufacturers and new industries and architects suggest new tools and new methods. Some are developed; some supplement various accessories or improvements and soon become indispensable for the driller. Others disappear after a few tests.

This discussion is therefore a picture of what currently exists because tomorrow, the genius of drillers will manufacture more efficient, more durable and more economical tools that will help carry out "boreholes" that are not feasible to construct at present.

The National Microtunnels Project, with more than 50 specific and highly specialized studies centered on boring machines or horizontal drilling and often in areas common to both techniques, has made progress in several fields and has currently gathered a large amount of information on state of the art processes.

This has led to the compilation of this text, which summarizes in a few pages all the recommendations for the use of techniques known as "horizontal drilling".

It is not possible to mention all those who have helped in accomplishing this National Project, particularly for the "horizontal drilling" part, but most of them are involved in the work at Workshop 7 of the FSTT. We thank them for their efforts in helping in the development of these techniques and making them familiar.

Chapter 10

Techniques and Principles of Operation for Horizontal Drilling

10.1. General information

The installation of pipes or cables can be planned by combining the trench and trenchless techniques, but the construction can also be carried out entirely by horizontal drilling; this is known as "line installation". This case also applies to sites varying from hundreds of meters to several kilometers.

Numerous sites have been constructed in France for the installation of gas pipes and to supply drinking water. The areas where horizontal drilling is used are in crossing roads or railway tracks.

In urban areas, it is used to cross intersections or shopping streets. However, this technology is often the only solution in cases where a river wider than 1 km needs to be crossed in order to install a gas pipeline with a diameter of 1,000 mm, for example.

The longest drilling carried out to date is of 2,000 meters and the largest diameter is 1,200 mm.

To make it easier to understand, the drill rigs are classified according to their pullback force. Every drill rig of 1 to 500 tons has its own application. They can either pull pipes of a maximum diameter of 1200 mm for a maximum length of 2 km or be used for simple crossings of road, rail or waterways networks.

The machines are thus classified according to their range of power (tension and torque).

Type of drills	Pullback force in kN	Maximum torque in kN.m	Mass in t
Mini	≤ 150	10 – 15	< 10
Midi	> from 150 to ≤ 400	15 – 30	10 – 25
Maxi	> from 400 to ≤ 2500	30 – 100	25 – 60
Mega	> 2500	> 100	> 60

Table 10.1. *Classification of machines*

The pipes used are generally in steel and in polyethylene. Sometimes PVC and cast iron are also used.

Figure 10.1. *Polyethylene pipeline in reels*

This technology is applicable to all sectors of public works and to any type of fluid transported.

Telecommunication: use of installed pipelines as ducts for the installation of fiber optic cables. Generally in these cases pipelines of small diameters from 25 to 50 mm are used. However, a pipe of up to 400 mm in diameter can be installed beforehand and then used ducts can be inserted if their number is quite large.

Electricity: use of installed pipes as ducts for the installation of electric cables. Generally in these cases, pipelines of medium diameters from 75 to 400 mm are used.

Water: direct use of pipelines for transporting drinking water.

Drainage: direct use of pipelines for transporting wastewater in the case of pumping.

Gas: direct use of pipelines for the distribution and transportation of gas.

Drains: direct use of pipelines for the collection of heap leachate at landfill dumps.

10.2. Different stages of horizontal drilling

All types of horizontal drilling have the following common functions:

– drilling of a pilot tube,
– monitoring and correcting the trajectory,
– reaming of the drilled borehole,
– installation of pipes or cables.

The equipment items can be differentiated according to their power (see section 10.1) and installation of drill rods.

The three stages for carrying out horizontal drilling are mentioned below: making a pilot hole and reaming (or the proportions) and the installation of pipes or cables.

10.2.1. *Pilot drilling*

A drill string is inserted into the ground applying on the bottom hole assembly a combined thrust and rotation action. This bottom hole assembly has a special feature of being asymmetrical in relation to the longitudinal axis. Mere thrust forces it to deviate, but rotation combined with the thrust gives it a straight trajectory.

This bottom hole assembly consists of an electronic tracking device and more or less sophisticated cutting tools. These tools are either simple drill blades or drill bits driven by hydraulic or pneumatic motors.

As the work progresses, rods of 0.5 to 6 meters in length are added according to the drill rig.

In addition, this drilling head is fitted with nozzles, which inject fluids that actively participate in the drilling process. Generally this fluid is a mixture of water, bentonite and additives all mixed in a tank and then injected under pressure into the rods.

In some cases, even foam (compressed air + additives) can be used.

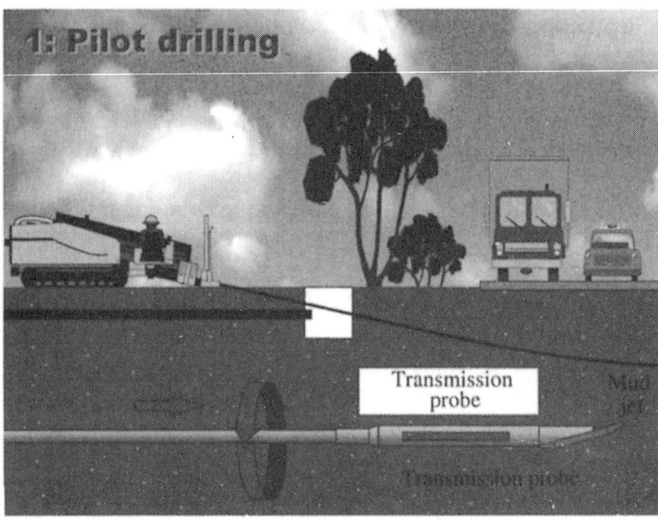

Figure 10.2. *Pilot drilling*

10.2.2. *Reaming*

Once the pilot hole has been made, the drill string comes out of the exit pit (most frequent case).

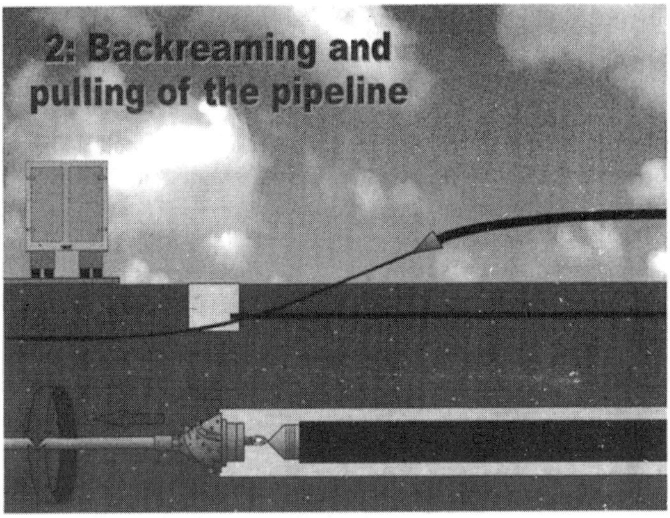

Figure 10.3. *Backreaming and pulling of the pipeline*

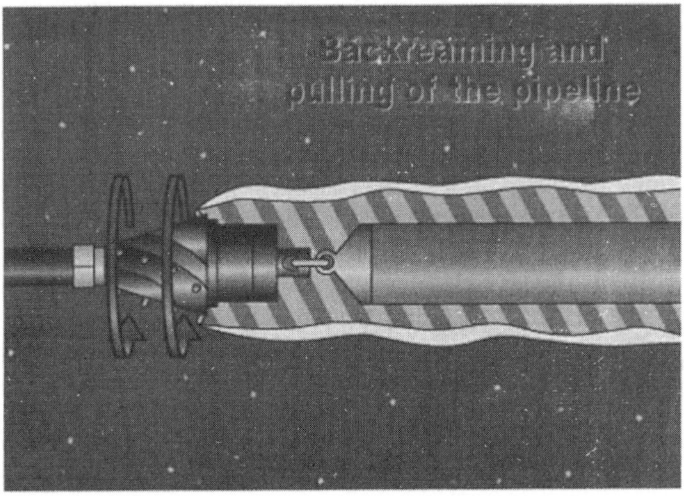

Figure 10.4. *Backreaming and pulling of the pipeline*

A reamer will replace the bottom hole assembly. Its function is to enlarge the previously drilled hole. This reamer is rotated and pulled by the drill string in the direction of the drill rig.

Generally, the drill string is towed behind the reamer. The hole is enlarged by successive stages of reaming of increasing sections until the desired section is reached (generally double the section required by the pipeline).

This reamer is equipped with injection jets for drilling "mud". This mud washes out and disposes of the cuttings created by the reamer, lubricates and cools the cutting tools and strengthens the borehole.

After the final reaming phase one end of the pipeline built earlier will be tied to the pulling head fastened to a suitable reamer. It will be pulled from one drilling end to the drill rig.

10.2.3. *Guidance and trajectory corrections*

The installation of pipelines by horizontal drilling must be able to:

1) constantly locate the position of the drilling head in the ground so as to respect the specified trajectory and avoid the already existing utilities: this is the *tracking* function;

2) know the pitch of this head and its direction to guide and divert its trajectory: this is the *guidance* function.

It is in fact the asymmetry of the head (wearing blade and nozzles in the case of a conventional head or a bend in the case of a mud motor) which by stopping the rotations of the rods diverts the trajectory, thereby correcting it.

The success of pilot drilling depends on the locating system, its accuracy and ease of use.

Two types of locating systems may be distinguished:

1) walk-over systems;
2) downhole systems or wireline steering systems.

10.2.3.1. *Walk-over systems*

These systems are suitable at most sites.

This technique is easy to implement, is safe for data transmission cables and has a low investment cost.

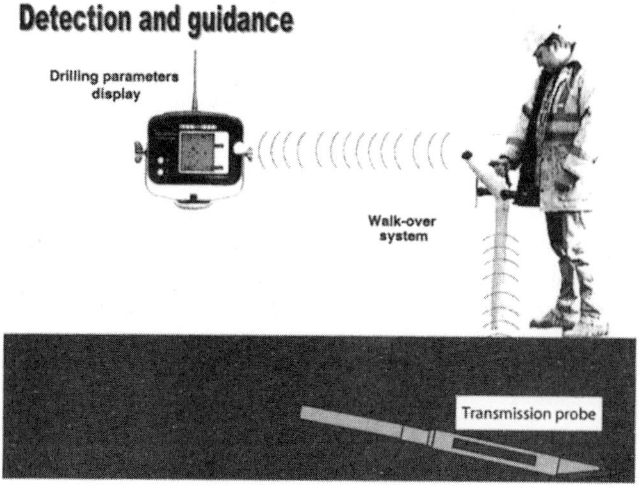

Figure 10.5. *Walk-over system*

Figure 10.6. *Detection and guidance*

The walk-over systems have disadvantages such as the reduction in accuracy of measurements increasing with the drilling depth, the influence of underground magnetic field interference and an operating range dependent on the life of the transmitter's batteries.

The systems consist of three elements:

– a transmission probe powered by batteries placed in the drilling head,

– a receiver that helps vertically locate the head and its direction,

– display of parameters (*remote*) on the drill rig.

10.2.3.2. *Downhole systems or wireline steering systems*

Used to cross obstacles such as streams or rivers, this guidance technique using data transmission cables has opposite advantages and disadvantages to the previous system with a more complex usage technique but with greater accuracy and range.

10.2.4. *Site organization*

10.2.4.1. *Administrative permits*

Some authorizations come under the project manager:

– traffic authorizations,

– authorizations related to labor laws,

– authorization related to environment laws,

– various authorizations.

Other authorizations come under the contracting authority or the owners of the property to be crossed:

– legal construction authorizations,

– authorizations granted to temporarily usufruct private property,

– regulations related to environmental matters.

10.2.4.2. *Access, site installation*

Access and installation drawings will be required depending on the different type of drilling equipment (weight and dimensions), lifting equipment, drilling fluid and movement of water containers.

These drawings will be subject to the existing traffic authorizations as well as encumbrance authorizations.

The pits for drilling mud will be made and secured before the drilling.

10.2.4.3. *Water*

Water is an essential element in making drilling mud.

It should be monitored to meet the drilling criteria (hardness, PH, salinity, etc.).

It will also be subject to local operational authorizations and authorizations related to environmental laws.

10.2.4.4. *Slurry transfers*

The reuse of excess cuttings and mud and the elimination of this mud must comply with the environmental laws.

10.2.4.5. *Work areas*

When preparing the site, the working premises must include the following parameters:

– type of drill rigs used, dimensions and supports,

– bore fluids: mixers, recyclers, water storage pond, area required by waste material,

– storage area for equipment and standby equipment (drilling rods, fluid and drilling products, etc.),

– assembling area for conduits to be installed,

– progress of conduits during pulling,

– workshop areas.

10.3. Different types of pipes or conduits

The pipes used for horizontal drilling are:

– thermoplastic pipes or conduits (PE, PVC and PP),

– metal pipes or conduits (ductile or steel casting).

Each has its field of use.

Field of use	PE	PVC	Steel	Casting
Water	X	X	X	X
Drainage	X	X	X	X
Gas	X	X		
Electricity	X			
Telecommunication	X	X		
Pollution removal drains	X			

Table 10.2. *Field of use of different types of pipelines in horizontal drilling*

10.3.1. *Thermoplastic pipelines*

Thermoplastic pipelines cover all the fields of use of horizontal drilling with a wide range of diameters and mechanical strengths. The use of thermoplastics ensures a suppleness and flexibility for optimum installation.

Three types of thermoplastics can be found: mainly polyethylene (PE), more rarely polyvinyl chloride (PVC) and polypropylene (PP).

Only pipes made of PE and PVC will be studied later in relation to dimensional characteristics, chemical mechanics, of packaging and implementation.

As regards the dimensions of the pipes for their use alone and not for the constraints of implementation by the horizontal drilling technique, the LAME formula, which expresses the relation between the internal pressure and the constraints inside the pipe, may be applied:

$$\tau = P \times ((D - e)/2\ e)$$

with:

- P = pressure inside the pipe in MPa,
- D = external diameter of the pipe in mm,
- e = thickness of the pipe in mm.

For a given material and pressure, the ratio of nominal dimensions of the tubing (diameter and thickness) is constant. This constant is represented by the acronym SDR that stands for the standardized dimensional ratio:

$$SDR = D/e$$

It is suitable to round off these values to the following numbers (according to the Renard series): 33 – 26 – 21 – 17 (or 17.6) – 13 (or 13.6) – 11 – 9 – 6.

10.3.1.1. *Polyethylene pipes*

A new material for some in the field of pipelines and yet quite old because it was introduced in 1933.

The development of these pipelines dates back to 1970 when they were first used for the distribution of gas.

Its mechanical characteristics make the material ideal for installation by horizontal drilling.

10.3.1.1.1. Dimensional characteristics

Depending on the field of application, the range of diameters available is different:

– *gas*: up to DN (nominal diameter) 400,
– *water*: up to DN (nominal diameter) 630 (or even beyond),
– *drainage*: up to DN (nominal diameter) 630 (even beyond),
– *electricity*: up to DN (nominal diameter) 400,
– *telecommunications*: up to DN (nominal diameter) 400,
– *drain:* up to DN (nominal diameter) 355.

10.3.1.1.2. Mechanical characteristics

Tables 10.3, 10.4 and 10.5 describe the mechanical characteristics according to the different types of polyethylene quality.

The general mechanical characteristics of polyethylene for PE 80 or PE 100 are indicated in Table 10.4.

The common range of pipes is indicated in Table 10.5 for PE 80 and PE 100.

Resin	MRS: minimum specific resistance (MPa)	Long-term hydrostatic stress σ (MPa)
PE 100	10	8
PE 80	8	6.3
PE 63	6.3	5
PE 40	4	3.2
PE 32	3.2	2.5

Table 10.3. *Mechanical characteristics according to different types of polyethylene*

	Unit	PE 80	PE 100
Density	kg/m^3	949 to 956	956 to 961
Melt index (190 °C, 5 kg)	g/10mn	0.2	0.15
Tensile strength	MPa	34	38
Ultimate elongation	%	> 600	> 600
Elasticity modulus	MPa	1,000	1,400
Brittleness temperature	°C	< – 100	< – 100
Linear expansion	K^{-1}	2×10^{-4}	2×10^{-4}
Electrical resistivity	Ω / cm	$> 10^{17}$	$> 10^{17}$

Table 10.4. *Mechanical characteristics of PE 80 and PE 100*

SDR	PE 80	PE 100
9	PN 16	–
11	PN 12.5	PN 16
13	PN 10	PN 12.5
17	–	PN 10
21	PN 6.3	–
26	–	PN 6.3
33	PN 4	–

Table 10.5. *Different ranges of PE 80 and PE 100*

Depending on the SDR, the minimum radius of curvature R_c for pipes is recommended below:

- SDR 9 to 11: $R_c \geq 25$ DN,
- SDR 13.6: $R_c \geq 30$ DN,
- SDR 17 to 26: $R_c \geq 35$ DN,
- SDR 33: $R_c \geq 40$ DN.

The acceptable pullback force for PEHD pipes (high density polyethylene) used under pressure is expressed in the following way, according to the ISO/TC 168 standard code of practice N 163 F:

$$\text{Pullback force} = (14 \times \Pi \times D^2)/(\text{SDR} \times 30) \text{ in dN}$$

For the usage of drill pipes, it has been shown that its values may be multiplied by two. However, care must be taken to consider the risks of ovalling, which can result from the shunting operation.

Furthermore, the temperature of the pipe considerably reduces the resistance to pullback.

Stresses and diameters	Force expressed in dN for the use of sleeves		
DN	**SDR 13.6**	**SDR 11**	**SDR 9**
90	1,700	2,100	2,600
110	2,600	3,200	3,900
125	3,300	4,100	5,000
160	5,500	6,800	8,300
180	6,900	8,600	10,500
200	8,600	10,600	13,000
225	10,900	13,400	16,400
250	13,400	16,600	20,300
280	16,900	20,800	25,500
315	21,300	26,400	32,300
355	27,100	33,500	41,000
400	34,400	42,600	52,100
450	43,600	53,900	65,900
500	53,800	66,600	81,400
560	67,600	83,500	102,100
630	85,500	105,700	129,300

Table 10.6. *Forces exerted on the PE pipes in Deca Newton*

To obtain the acceptable pullback force in tons, all that is needed is to divide the value in Table 10.6 by 1,000.

FOR EXAMPLE: for a DN160 SDR11 tube, the force acceptable in tons is:

6,800/1,000 = 6.8 tons

NOTE: when the pipes are used to carry fluids under pressure (water and particularly gas), the pullback during installation is a parameter which affects its ageing. The concerned companies stipulate the pullback threshold values (for safety reasons, these values are often less than the resistance values of PE). In fact, the frictional forces of the reamer and train of rods absorb a part of the force measured.

10.3.1.1.3. Chemical characteristics

Polyethylene is chemically inert within its temperature range of use for all practical purposes.

When the pipes have to be installed in internal or external environments where the concentration of certain chemical products are very significant, it is recommended that the NF T 54-070 and ISO/TR 10358 standard tables are referred to.

10.3.1.1.4. Packaging

There are three types of packaging possible and adaptable to the condition of the site. These reduce collapses and enable the company to optimize its costs:

- spool: 25, 50 or 100 m (up to DN 63),
- reel up to DN 160,
- rod: 6 or 12 m.

10.3.1.1.5. Implementation

The choice of pipes is essential for the proper execution of the work.

Generally, SDR 11 pipes are recommended because they offer good resistance to tensile stresses during installation.

For smaller sites having no particular difficulties, SDR 13.6 pipes may be used.

In case of high tensile stresses, PE 100 pipes are recommended. There are two types of joining methods: either by welding or by mechanical type assembly (screwing down, etc.).

Butt fusion

Butt fusion by heating element is used to join PE pipes of similar thickness and compatible melt indexes.

This procedure consists of raising the ends of the pipes to the welding temperature using a heating plate (mirror).

A good butt fusion done with skill will maintain the continuity of the pipeline perfectly with an identical mechanical strength.

Figure 10.7. *"Mirror" welding of PE*

Examples of the end-to-end welding parameters are given in Table 10.7.

Wall thickness (mm)	Preheating thickness of the rim at the end of the heating (mm)	Heating Heating duration (seconds)	Removal of the electrically heated plate Maximum duration (seconds)	Duration of the application of pressure until the set point (seconds)	Cooling time under assembling pressure (minutes)
4-5	0.5	30-70	3-5	3-6	3-6
5-7	1	70-120	4-6	4-8	6-10
7-12	1.5	120-190	5-8	8-12	10-16
12-19	2	190-250	6-10	10-15	16-24
19-26	2.5	250-330	7-14	15-20	24-32
26-37	3	330-460	8-17	20-25	32-40
37-50	3.5	460-600	7-20	26-35	40-45

Table 10.7. *Welding characteristics of PE*

NOTE: automatic equipment with programmed fusion phases exists according to the characteristics of the pipe.

Assembly by screwing down

Mechanical joining by screwing down helps respond to the restrictions of congested urban sites and avoids the use of welding procedures.

The fields of application of polyethylene are varied:
- pipes for gas (pipes with yellow bands),
- pipes for drinking water (pipes with blue bands),
- electrical cable ducts (pipes with red bands),
- telecommunication ducts (pipes with green bands),
- pipes for pressurized drainage (pipes with brown bands),
- depollution drains (pipe colored according to the field of use).

10.3.1.2. *Polyvinylchloride pipes*

PVC pipes have been used for more than half a century but their use in the field of horizontal drilling remains less common than vertical drilling (they are mainly used in drilling for water).

10.3.1.2.1. Dimensional characteristics

The range of PVC pipes for the implementation of horizontal drilling extends from DN 110 to DN 400. The most commonly used diameters are: 110, 125, 140, 180, 200, 225, 250, 315 and 400. For every diameter numerous thicknesses are available.

10.3.1.2.2. Mechanical characteristics

The main mechanical properties of PVC pipes are indicated in Table 10.8.

Density (kg/m^3)	1,300 to 1,400
Tensile strength (MPa)	45
Ultimate elongation (%)	80 to 200
Elasticity modulus (MPa)	2,400
Linear expansion (K^{-1})	1×10^{-4}
Electrical resistivity (Ω / cm)	10^{16}

Table 10.8. *Mechanical properties of PVC pipes*

10.3.1.2.3. Chemical characteristics

PVC is known for its remarkable resistance properties against various chemical agents, which makes it insensitive to corrosion.

It is particularly resistant to:

– various atmospheric agents,

– H_2S (hydrogen sulphide),

– alcohols and aliphatic hydrocarbons,

– oils.

10.3.1.2.4. Packaging

Considering the high rigidity of PVC pipes, packaging in rods is the only possibility.

They are generally of pre-coupled lengths of 4 or 6 meters that have to be joined.

10.3.1.2.5. Implementation

Joining is almost always done by gluing.

The first stage consists of using a scouring agent to properly clean and degrease the part to be stuck.

Next, the use of a strong solvent adhesive is recommended to stick the two parts together. These glues generally have a long gluing time (5 to 15 mins approximately). During the gluing time, no mechanical stress must be applied to the junction.

Furthermore, one must always respect the drying time of the adhesive mentioned by the supplier.

The amount of adhesive and scouring agent varies according to the external diameter of the pipe (see Table 10.9).

	Number of liters required for 100 joints	
DN	Adhesive	Scouring agent
110	2	0.5
125	3	1
140	4	1.5
180	6	2.5
200	9	3
225	10	3.5
250	12	4
315	20	6.5
400	30	10

Table 10.9. *Amount of adhesive required depending on the diameter of the PVC pipes*

There also exist a variety of threaded pipes for the same range of diameters where the external diameter remains constant (without flaring at conduit couplings).

The fields of application of PVC pipes are:

– telecommunications,

– drilling for water.

10.3.2. *Metal pipelines*

The use of metal pipelines for horizontal drilling began with the use of steel to transport gas and thus increase the speed of installation during difficult passages of feeders (crossing a river, canal, etc.). In 1995, manufacturers of ductile cast iron pipelines proposed a solution for AEP (drinking water supply) and drainage.

10.3.2.1. *Steel pipes*

Steel pipes were used to transport gas and hydrocarbons as well as for water and drainage. Steel pipes offer very high resistance. They retain the same external diameter at the joints by welding.

These welding areas are externally protected from corrosion by sleeve coatings. Special care must be taken to avoid damaging them and the entire exterior covering during pulling of the pipeline.

The choice of the covering will be dealt with in a specific study so as to respond to the resistance to corrosion. Steel is the most commonly used type of pipeline (90%).

10.3.2.1.1. Dimensional characteristics

Steel pipes are predominantly of large diameters.

Depending on the manufacturing processes, the external diameters can range from 60 inches (approximately 1,700 mm) to more than 112 inches (approximately 3,000 mm) with spiral welding.

External diameter (mm)	**Thickness (mm)**	**Type of welding**
540 to 1700	7 to 40	Lengthwise welding
430 to 3,000	4.5 to 20	Spiral welding

Table 10.10. *Type of welding depending on the dimensional characteristics of the steel pipes*

Generally, in the field of horizontal drilling, the external diameter rarely exceeds 1,200 mm.

10.3.2.1.2. Physical characteristics

The main mechanical properties of steel pipelines are given in Table 10.11.

Yield strength (MPa)	255
Tensile strength (MPa)	410
Ultimate elongation (%)	20
Elasticity modulus (MPa)	20,000
Thermal expansion (m/m°C)	$12.4\ 10^{-6}$

Table 10.11. *Mechanical properties of steel pipelines*

10.3.2.1.3. Chemical characteristics

The pipes are covered with a coating to protect against corrosion:

– an external coating in polyethylene (PE) or polypropylene (PP),

– an internal coating in cement mortar or epoxy.

10.3.2.1.4. Packaging

Packaging is done exclusively in rods with lengths that may reach 18.3 meters.

10.3.2.1.5. Implementation

Assembly by welding is the best joining system for steel pipelines. This technique ensures a perfect seal as well as the continuity of mechanical strength.

It requires specialized skilled labor and strict onsite monitoring. Furthermore, before the arc welding procedure, the anticorrosion protection must be removed from the ends, and these should be restored carefully after joining so as to prevent the pipe from corroding.

The fields of application of steel pipelines are:

– mainly gas,

– drinking water supply,

– drainage,

– ducts for cables and pipelines.

10.3.2.2. *Pipes in ductile cast iron*

The use of ductile cast iron pipelines to carry out horizontal drilling dates back to about a decade in the French market.

It consists of pipelines equipped with a TYS-K type interlocking system or standard Ve (Ve: external interlocking).

NOTE: a pulling head was specially designed for this application, which enables hooking of the first pipe.

Compared to steel, the advantage of cast iron is its high resistance to corrosion. Its disadvantage is in excrescence at the joints. This disadvantage requires an increase of the backreaming section.

For a pipeline in cast iron, the backreaming often corresponds to four times the cross-section of the pipeline (twice for steel).

10.3.2.2.1. Dimensional characteristics

The range of ductile cast iron pipelines is from DN 100 to 500 mm and this is for drinking water supply or drainage.

DN	Maximum external diameter of the pipe-puller (mm)	Maximum external diameter of the pipe juncture (mm)
100	186	182
150	241	236
200	297	293
250	357	348
300	415	406
350 to 500	Consult the suppliers	

Table 10.12. *Backreaming of the tunnel depending on the diameter of the cast iron pipelines*

An angular deviation of 3° is permissible per joint, which adds up to a deviation of 32 mm per pipe whose length is of the order of 6 meters (i.e. a minimum drilling radius of curvature of 360 meters).

The characteristics of the backreaming diameter of the borehole are indicated in Table 10.12.

The minimum internal drilling diameters are given in Table 10.13 as a guide.

DN	Aligned installation (mm)	Installation with deviation (mm)
100	200	225
150	255	280
200	310	325
250	370	385
300	430	445
350 to 500	Consult the supplier	

Table 10.13. *Backreaming of the borehole depending on the internal diameter of the pipelines in cast iron*

The acceptable operating pressures are given in Table 10.14.

DN	PFA (MPa)	PMA (MPa)	PEA (MPa)
100 to 150	2.5	3.0	3.5
200 to 300	1.6	1.9	2.4
350 to 500	Consult the supplier		

Table 10.14. *Operating pressures for cast iron pipelines*

For the definition of PFA, PMA and PEA pressures, refer to the NF EN 545 standard. The laying length of the pipes is 5.97 m.

10.3.2.2.2. Physical characteristics

Ductile cast iron is different from the traditional grey cast iron due to its mechanical properties (flexibility, resistance to impacts, elongation, etc.). This is due to the spheroidal shape of the graphite particles.

The mechanical properties of ductile cast iron are indicated in Table 10.15.

Yield strength (MPa)	270
Tensile strength (MPa)	420
Ultimate elongation (%)	12
Elasticity modulus (MPa)	170,000

Table 10.15. *Mechanical properties of cast iron pipelines*

The maximum tensile stresses are given in Table 10.16.

DN	Type of joint	Maximum tensile stress (daN)
100	TIS K	4,000
150	TIS K	8,000
200	TIS K	11,000
250	TIS K	16,000
300	TIS K	20,000
350	UNI Ve	24,000
400	UNI Ve	30,000
450	UNI Ve	37,000
500	UNI Ve	46,000

Table 10.16. *Maximum tensile stresses for cast iron pipelines*

10.3.2.2.3. Chemical characteristics

To meet the contact requirements with bentonite used for drilling, the pipelines in ductile cast iron are externally coated with a thick covering of polyethylene, which protects the pipe from scratches while ensuring smooth entry into the ground. This covering adds a thickness of 2 mm on the casing of the pipeline and a sleeve in elastomere protects the junction. These protections are suitable for operating temperatures of fluids not exceeding 30°C.

10.3.2.2.4. Packaging

Pipelines with DN lower than or equal to 300 are packed in loads ranging from 4 to 15 pipes. On the other hand, pipelines with DN over 300 mm are not packaged.

10.3.2.2.5. Implementation

Here are some simple rules that need to be followed for the installation to ensure a proper positioning of the interlocking system:

– check the presence and compliance of the bevel at the joining ends,

– check the cleanliness of the seal ring housing,

– lubricate the bevel properly, the connected ends of the pipe and the exposed face of the seal ring,

– assemble the pipes in a well-aligned manner,

– check the proper positioning of the seal ring after fitting.

The fields of application of pipelines in ductile cast iron are:

– mainly for gas,

– drinking water supply (blue color pipe),

– drainage (red color pipe).

Chapter 11

Summary of Parameters Affecting the Start of a Building Site

11.1. Summary of parameters affecting the execution of horizontal drilling

1) nature of the ground,
2) groundwater and soil permeability,
3) presence of obstacles if any,
4) type of pipe or the pipes to be installed,
5) length of the drive,
6) radius of the curvature,
7) characteristics of the mud and incidence of lubrication,
8) characteristics of the rig,
9) regularity of the profile, steering and guiding,
10) preliminary negligence,
11) surroundings of the site and its environment,
12) execution (details) timing or timeframe,
13) climatic conditions.

11.2. Parameters related to the ground

The nature and the mechanical properties of the soil are generally the first elements taken into account in a feasibility study of a horizontal drilling project.

Therefore, it is necessary that all the participants in the project use:

– the same classification,

– the same terminology.

Type of soil	Normal classification	Impact on the project
Class A	Fine ground: silt to clay	Joining, stuffing with occasional clogging
Class B	Sandy or gravely grounds with fine sand: sand-gravel more or less clay	Abrasiveness, deviation and eventually sinking followed by clogging
Class C	Grounds comprising of fines and the main elements: clay or chalk with flint, fill or moraines	Abrasiveness, deviation and eventually sinking followed by clogging
Class D	Ground that is not sensitive to water: sand and pure gravel	Partial to total mud loss Instability of pulverulent materials
Class R	Rocks: carbonated, clay, silica, saline, magmatic and metamorphic	Rapid wear and tear of tools and the rig Contamination by salts in evaporation Impossibility of digging the tunnel across grounds/rocks Problems of fractured zones – mud loss – local instability
Specific materials	Organic ground, industrial wastes	Mud leakages, contamination by organic matter

Table 11.1. *Classification of soils and the impact on horizontal (digging) drilling*

One possibility is to use the French standard NF P 11-300 of September 1992 – classification of the materials that can be used in the construction of embankments and the layers of infrastructures related to roads.

This is the latest classification available, even if it is insufficient for underground work.

It can be usefully completed by the AFTES classification, the French committee for (trenchless) underground work.

It is thus possible to compile a table of summarizing the ground impact on the project, as per the classification of standard NF P 11-300 (see Table 11.1).

11.3. Parameters related to groundwater and soil permeability

The chemical composition of the groundwater can affect the physio-chemical characteristics of the mud and therefore its rheological properties or its filtration.

A chemical analysis of the water, the piezometric level, the circulation speed and the soil permeability are data needed for the driller.

11.4. Parameters related to obstacles

Obstacles of huge dimensions, natural or artificial cavities, buried structures (old walls, passages), foundations or archaeological ruins can call the method of execution to be questioned.

The crossing of voids may require specific methods, such as filling, injections or congealing. Extracting them is also a possible solution. They must be located by the usual methods used in ground recognition.

11.5. Parameters related to the nature of the pipeline to be installed

Installations of a steel pipe for gas or a PE tube for the gravity sewerage are tackled in a different way. Frictions and the number of reamings for installing a gas passage of several hundred meters are some of the primordial elements.

For the sewerage pipe, the accuracy of the guiding system will be important in order to take account of the slight sloping of gravity. Of course, the number of pipes or caves to be installed is yet another factor which influences the drilling to be undertaken.

11.6. Parameters related to the drive length

The preceding section specifies the importance of the length of the drive that is to be made.

The longer the drive, the more that the capacity of the drill will be determined to maintain the traction efforts necessary due to friction constraints.

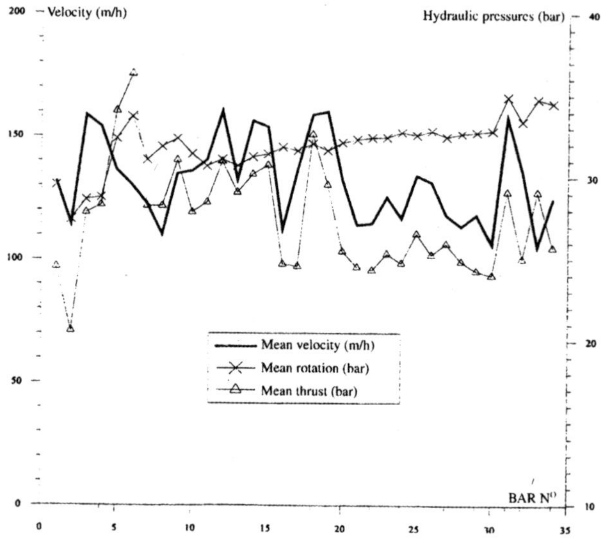

Figure 11.1. *Diagram of factors (registered) recorded in the course of the drilling the pilot hole*

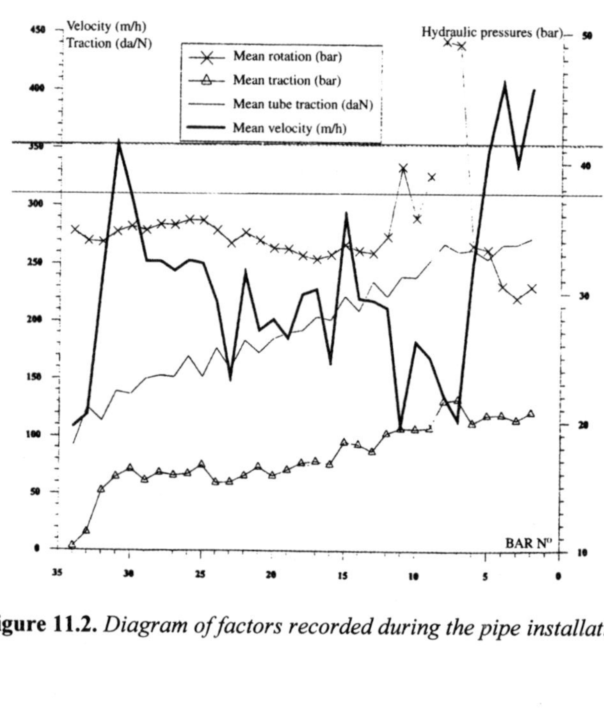

Figure 11.2. *Diagram of factors recorded during the pipe installation*

11.7. Parameters related to the radius of curvature

The nature of the soil (the different layers in particular), the topography, the type of pipe and the section of drill rods constitute some of the parameters influencing the choice of the radius of the curvature as well as its feasibility.

11.8. Parameters related to the characteristics of the drilling mud

This factor is essential for the drilling to be successful. Its nature, quality and method of employment are essential for the correct installation of the pilot bore, the reamings and for the drawing of the pipes.

11.9. Parameters related to the characteristics of the drilling rig

Considering that the length of the driver is affected by the pipe diameter and the nature of the ground to be covered, it is essential to choose the power of the machine carefully.

11.10. Parameters related to the regularity of the profile, the piloting and the guidance

The precision of the drive directly affects the friction stresses on the pipe and also the length of the driver and any eventual perforations in the pipe in case of abrasive or rocky soils.

11.11. Parameters related to preliminary exploration

The following chapter will address the importance and the necessity of preliminary exploration in a drilling project.

Exploration of the physical obstacles (pipelines, blocks) is fundamental for the success of a drilling project. Exploration with a tool which is not suitable can bring about unforeseen delays in the progress of the project.

11.12. Parameters related to the (overall dimensions) congestion of the site

Obstacles at the site can cause some recoil on the trajectory to be achieved and can affect the radius of the curvature at entry as well as exit. In fact, tube installation can sometimes prove to be difficult due to space constraints.

11.13. Parameters related to delays

A shortened timeframe is often an important factor in understanding the different stages of drilling. It is therefore essential for the planner behind the project as well as the driller to keep this in mind.

11.14. Parameters related to weather conditions

For this type of building site, considering that water will be used, it is important to drill at temperatures higher than 0°C.

Chapter 12

Guidelines for Explorations

12.1. General theory of explorations

12.1.1. *General objectives*

Geotechnical explorations prior to horizontal drilling projects are considered here in a broad sense: they include research of geological, hydrogeological and geotechnical data, as well as hidden natural or artificial obstacles, which can interfere with the drill path.

They have four objectives:

– to contribute to the optimization of the geometry of the project drill path plan, longitudinal profile, number and installation of pits), knowing that many more constraints will have to be taken into account;

– to optimize the choice of the drill and to size the tubes to be introduced;

– to identify the major hazards which can hinder the success of the drilling (hardness of the ground and wear and tear of the tools; clay grounds may increase friction and permeable soils may affect the efficiency of the drilling fluids);

– to outline an implementation method and enable the project manager and companies to estimate the cost and construction time.

These explorations have to be carried out at the earliest stages of the project because it is then that the chances of saving money are greatest, in view of the first objective in particular. The risk of carrying out explorations "in vain" less significant the savings that can be identified during the design stage of the project itself.

An implicit objective of explorations is also to make sure that the planned layout does not present any insurmountable obstacles which will make the project impossible; and even if there are any obstacles the earlier these are known, the better.

12.1.2. *Stages of explorations*

The normal progress of the explorations includes four successive stages, which will be detailed in the following chapters:

– documentary survey,

– geophysical investigation,

– geotechnical boreholes and tests (*in situ* and in the laboratory),

– summary of all the explorations, and preparation of a geotechnical dossier which will be attached to DCE [dossier de consultation des enterprises (company reports)] in the case of public works contract or to the private contract proposal of the works.

To carry out these explorations, the usual recommendations relating to all geotechnical projects remain valid:

– first look to locate the project in a regional and local geological context, in order to be sure of fully understanding the configuration of geological layers in the path of the project; this is even more so the case for hydrogeological conditions;

– plan an investigation program consisting of, preferably, two stages separated by an intermediate summary, which will help "rectify the drive" in view of the initial results;

– reserve about 20% of the available budget for the explorations or tests not allocated in advance, in order to have resources to react immediately in case of unforeseen conditions;

– encourage the intending contractors to come and observe the core samples and study the detailed results of the explorations not attached to the DCE.

12.1.3. *Cost of explorations*

Unlike trenching work where one can adapt to the ground encountered on a day-to-day basis, the work with horizontal drilling requires particularly careful explorations (which are therefore relatively costly), for several reasons:

– the conclusions drawn from explorations will be irreversible: once the machine has entered the ground, changing or modifying it is detrimental to the success of the project;

– the economy of the project, for the company as well as for the contracting authority, is incompatible with a serious overestimation of the penetration speed;

– the possible damage to networks that are not investigated or cases of accidental pollution can lead to increase in costs, penalties or unacceptable reinstatement durations.

For a horizontal drilling project, it is normal that exploration costs are proportionately greater than for a man-accessible tunnel, and all the more so as the length reduces: the exploration/work ratio is normally about 5 to 15%.

This initial investment, designed to reduce the burden of additional costs, must be regarded as an integral part of the project and must consequently be planned and budgeted for. Thus, for a 200 m drilling, an investigation budget in the region of €3,000 is not too excessive.

These Figures can be compared to the recommendations of the *U.S. National Committee for Tunneling Technology,* which recommends an average 0.5 m of boreholes per linear meter of the tunnel. This criterion for length must, of course, be weighted according to the geological variability and uncertainties peculiar to the site under consideration.

Of course, it is up to the contracting authority to bear the cost of these initial explorations: they alone have the resources and above all the available time to carry them out correctly.

Furthermore, in case of insufficient explorations, it is them who will bear the consequences in terms of cost and time; similarly, it is they who will profit from the information obtained thanks to a thorough investigation.

12.2. Data to be acquired

12.2.1. *Geological configuration of the site*

Understanding the geological configuration of the site, which generally means obtaining a longitudinal geological cross-section, is the fundamental objective and the basis of the rest of this process. This understanding can result only from *local data set in a regional context*, this latter enabling us to interpret and then extrapolate them conclusively.

It is only after having delimited and prioritized these geological units that we can try and characterize each of them from a physical and mechanical perspective.

As an example, it is only with a good knowledge of the regional geology that we will be able to know if there is a risk of encountering erratic blocks in the moraines, or large stones at the bottom of alluvia, or flint embedded in the chalk, etc. – all things that boreholes have very little chance of detecting directly.

Similarly, when boreholes encounter a substratum at varying depths, the designation of the top surface (and the estimation of the uncertainty associated with this designation) can only be conclusively done on the basis of a known regional model of erosion or weathering.

The explorations must also endeavor to detect and characterize a certain number of objects or geological configurations that are particularly problematic for drilling, notably:

– formations whose constituents are *mechanically heterogeneous* (anthropic backfills, moraines, burrstone clay, slope scree, irregularly weathered rocks, flint embedded in a softer matrix ,etc.);

– soils with *highly coarse granulometry*, such as torrential alluvia: the presence of gravel (2 to 20 mm) or shingle (20 to 200 mm) may make it difficult or even impossible to carry out the pilot drill by creating difficulties in steering the drill, eventual jamming of the drill string, instability of the tunnel and the loss of the drilling fluids (Figure 12.1);

– *plastic clay*, which can cause formidable problems of swelling or sticking in the presence of water;

– *soft soil*, in general, which causes steering problems;

– *interfaces between layers*, as they can be much more difficult to drill through with a boring machine than the separate grounds individually; moreover every effort is made to locate the longitudinal profile to prevent such a situation, all the more so as it is generally impossible to locate the interfaces in advance to the nearest decimeter ;

– finally, *natural cavities*, either of karstic (in gypsum or limestone) origin or caused by the washing out of fine materials in non-saturated zones.

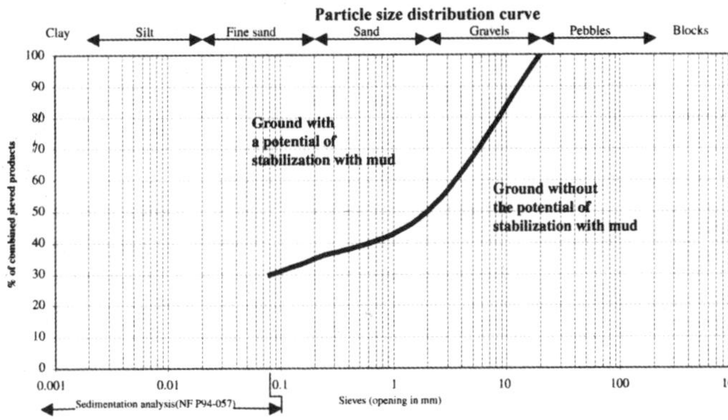

Figure 12.1 *Critical particle size distribution curve*

12.2.2. *Hydrogeological conditions*

It is essential to know the probable and maximum level of the water table during the works, to ascertain the penetration rate of the machine. Two secondary parameters may also be useful:

– the *chemical composition* of the underground water as well as its pH (particularly when it is rich in sulphates), for possible interactions with the drilling mud, for problems of sticking as well as for its aggressiveness in relation to the product pipe;

– the horizontal *speed of flow* of the water table, which, if very high, can endanger the stability of the drill hole.

12.2.3. *Geotechnical characteristics of the soils*

The parameters useful for the development of a drill project are indicated in Table 12.1, where we have distinguished between the mandatory parameters to be measured in all cases, and the additional parameters, which are generally to be estimated indirectly.

	Mandatory parameters	Additional parameters
Physical characteristics	Particle size distribution Plasticity Blue value Specific weight, water content	Mineralogy Ability to stick, pH
Mechanical characteristics	Strength deformation properties, Abrasiveness and hardness	Creep and swelling parameters Dilatancy
Hydrogeological characteristics	Average and maximum level of the water table	Permeability Flow velocity

Table 12.1. *Useful geotechnical parameters for drilling projects*

12.2.4. *Pockets and artificial obstacles*

This aspect of explorations gains importance as the drill profile is shallower (depth < 5 m), and the length of time the site has been urbanized.

The following obstacles must be investigated, as they are the most challenging for a horizontal drilling:

– demolition and landfill products (concrete, scrap metal, etc.),

– old foundations (masonry work, piles, etc.) of buildings that have been subsequently demolished and foundations of existing works,

– old wells (originally, there was one in the courtyard of every house),

– forgotten cellars, quarries or underground shelters that have often not been filled in,

– various pipelines and cables, in operation or abandoned; these networks are more often than not located in the 0–3 m section, but their position is never completely in keeping with the one that is indicated on the drawings, if they exist, etc.

12.2.5. *Environmental parameters*

The importance of these parameters related to the geotechnical conditions, is extremely variable depending on the situations. In all cases, it is appropriate to examine at least the following points at the very least:

– *initial condition of the pollution of the ground*, concerning possible interactions with the drilling mud and its additives, as well as the constraints that may result from this pollution by the stockpiling of earth (evacuation to a storage area considerably increases the cost of the structure);

– *evacuation of earth*: identification of possible locations for stockpiling, considering a possible initial pollution and the drilling fluids used;

– *condition of the existing frame* and neighboring underground structures, from the perspective of their sensitivity to possible movements caused by the horizontal drilling (compaction and uplifting);

– *construction of access shafts*: the locations and surfaces available, accessibility, sensitivity of the site to nuisances (noise, soil, etc.).

12.3. Methodology and means of explorations

The means to acquire the data listed in section 12.2 include desk, site and laboratory studies.

12.3.1. *Documentary survey*

The consultation of previous documents is essential, particularly for urban areas: there is always data of geotechnical interest to be found (drilling databases, geological maps, files of earlier structures, old drawings, etc.).

The issue is all the more important as new explorations risk being difficult to carry out. This consultation will also provide additional indications that are indispensable for two reasons:

– the presence of *existing or forgotten networks* and underground obstacles; the first are noted (more or less accurately, etc.) on drawings of statutory companies or in databases of some towns;

– the historic levels attained by the *water table* in the past; we must note, in fact, that water levels measured during a drilling campaign will be subject to all sorts of influences (rain, rise in water level, blockage in neighboring catchment points, etc.), which will not necessarily be identical during drilling work.

At this stage a visit to the site, excavations or earthworks situated in an identical or similar geological context should be envisaged, considering that a “good” outcrop is more important than several drilling operations, even if it is not exactly part of the plan.

12.3.2. *Geophysical investigations*

12.3.2.1. *Objectives*

The use of geophysical methods has many advantages for the investigation of horizontal drilling projects:

– supply of a 2D image on the distribution of the soil along the projected line route (most methods providing continuous profiles);

– optimizing the layout of exploratory boring (with or without boring cores);

– possibility of laterally extrapolating the boring data, once the geophysical profiles have been tested on these borings;

– finally, emphasizing the localized heterogeneities, which are not likely to be encountered by boreholes (which assumes the use of a geophysical method that is well suited to the nature of the "objects" sought and their host).

These various applications, whose order is not irrelevant, show that it is preferable to begin *geophysical explorations* before boring.

12.3.2.2. *Advantage of various methods*

Table 12.2 summarizes the main geophysical methods used during the investigation of drilling projects with their advantages and drawbacks; only the most common methods have been listed, without prejudging new methods or combinations of methods currently being developed.

The most widespread method is the *geological radar*.

Its main interest is to detect and locate objects of known nature but whose position is unknown (typically, networks and horizontal or vertical interfaces), rather than to identify discontinuities or unknown objects.

It is also used to "indicate" the presence of heterogeneities, with an excellent resolution (5 to 10 cm at less than 5 m depth) but their exact nature will have to be specified by other means of exploration.

Unfortunately, the radar is not very effective in silt-laden ground, and blind in humid clayey ground. Very precise calibration tests were carried out in 1997 at the experimental LCPC site at Nantes, which helped prepare a "catalogue of radar signatures" for the most common obstacles (see report no RS 17 of the National Microtunnels Project).

Method **Basic principle**	**AREAS OF APPLICATION** **Adv. = main advantages**	**DISADVANTAGES**
Geological radar Reflection of electromagnetic waves on the interfaces	Detection of interfaces, various networks and obstacles (metallic or non-metallic) Adv.: rapid, not very cumbersome Continuous profile at high resolution	Tricky implementation and interpretation (by specialists) Blind in clayey ground or in the water table. max. depth 5–10 m. Penetration < 2 m if networks Ø 20 mm
RMT (radio-magnetotellurgy) Measurement of the resistivity via perturbations in the electromagnetic field of a radio transmitter	Geological identification of ground and buried obstacles (metallic or non-metallic). Adv.: continuous profiles, very rapid, cheap, Good lateral resolution. No geological negative guidance	Investigated depth not controlled well Disrupted by metallic networks, but this can be an advantage! Not very suitable for urban areas (strongly interfered signals)
Electromagnetic methods with close transmitter Creation of Foucault currents, measurement of the induced field	Geological identification of the ground Detection and location of metallic networks Adv. Easy to implement and efficient	Penetration depth: – EM 31: 3–4 m, – EM 34: 10 m, Discontinuous profiles Frequent interference in urban areas
Electrical method Measurement of apparent resistivity by injection of direct current and measurement of the potential difference	Geological identification of the ground Adv.: control of the depth and the lateral and vertical resolution Suitable for all types of ground	Discontinuous profiles In urban areas, ensure good electrical contact with the soil
Electrostatic quadripole Same as electrical method (current 10 to 50 kHz)	Same as electrical method, but: - much quicker, continuous profiles, - can be used for road surfaces	Limited experience
Electrical method in aquatic sites Measuring the resistivity by current injection; measuring ΔP by electrodes dragged at the bottom of the water	Geological identification of the ground Adv.: control of the depth and the lateral and vertical resolution Continuous profiles. Low cost, suitable for all types of ground	

Seismic refraction Refraction of seismic waves on layers at speed increasing with depth	Looking for the dimension of a substratum Assessment of the mechanical characteristics of layers	Assumes a low dip Poor horizontal resolution In town: frequent static; requires lightweight sources
Seismic reflection Reflection of seismic waves on contrasting interfaces	Definition of the geometry of layers, in soil or in aquatic site Adv.: rapid method, providing a continuous profile	In town, preferable to work at night Under water, blind if the bottom is muddy with gas bubbles
Seismic surface waves (SASW) Analysis of the dispersion of seismic surface waves	Identification of hard or decompressed spots (based on the distribution of shear modulus)	Assumes stratified ground Can be used in town (light sources) Not widely used in France
Microgravimetry Local variations of the gravitational field	Localized search for spaces, decompressed areas and rippling of the substratum	Poor resolution; Small cavities (Ø < 2 m) Not visible if depth > 5 m Low efficiency. Requires complex corrections in urban areas

Table 12.2. *Areas of use and limitations of geophysical methods*

Figure 12.2. *Geological radar*

After the radar, several geophysical methods with large efficiency must be listed:

– *radio-magnetotelluric* (RMT), a method that is especially suitable for unused sites and which provides continuous profiles, such as radar;

– *electrical prospecting* by electrostatic quadripole (also continuous profiles);

– electromagnetic prospection (type EM 31 or EM 34).

These very rapid methods are better suited than radar to describe the distribution of soils and their nature, but they are less effective in detecting and especially locating precisely obstacles and horizontal interfaces.

The seismic methods also need to be mentioned, in spite of their difficulties for use in urban areas, in particular:

– *seismic refraction*, which gives very good results in geological configurations characterized by a series of layers of increasing velocity with the depth, even under the water plans of low depth,

– *high resolution seismic reflection*, which makes it possible to point out the potential existence of "reflectors" linked to contrasts in density and/or rigidity of layers, and which is very well adapted for the study of crossing water courses; in the case of water plans of low (<10 m), it is recommended that seismic refraction or *electrical prospecting by direct current be used.*

Finally, *microgravimetry* needs to be mentioned: it is a localized method used to detect the presence of cavities whatever the nature of the host ground; but this method is effective only for cavities with low depth (a 2 m Ø gallery remains invisible if its roof is at more than 5 m).

It is quite expensive as it requires measuring stations that are very close together, particularly if we are looking for small cavities.

12.3.2.3. *General recommendations*

In all cases, one must first endeavor to design a projected geological section based on the documentation, to estimate its uncertainty and then to consult specialists to select the most suitable geophysical method(s).

For a drilling project situated at a depth of less than 5 m at the town centre, where the geological structure is largely known, it is recommended to start with the following geophysical methods:

– geological radar, except in the case of clayey ground or the surface water table;

– in other cases, electrical or electromagnetic prospection EM 31 (to be replaced by EM 34 for a project located at a depth of between 5 and 10 m).

For a project in a ground that is less obstructed (outside the town centre), first try to clarify the geological structure over the entire planned course by using RMT, or seismic methods if the ground is suited to these. The detection of isolated obstacles will be done subsequently, by geological radar or electromagnetism.

Microgravimetry will be used more frequently only to precisely locate the surface cavities whose existence is proven, but whose exact position is not known.

Finally, it must be indicated that the main French operators have signed a charter called the "Code of good practices in geophysics", which must be taken into account in the specifications.

12.3.3. *Drilling and* in situ *geotechnical tests*

12.3.3.1. *Test drilling objectives*

Exploratory boring is indispensable as it meets several needs:

– to precisely define the geological section of the path,

– to measure the level of the water table and possibly the permeability of the ground,

– to take samples for geotechnical tests at the laboratory,

– to possibly carry out geotechnical tests *in situ*.

In general, where there are no observable outcrops, at least one boring must allow a *direct visual description* of the nature of the ground to be made and samples to be taken.

Only cored boreholes and pits dug with a mechanical digger fulfill this objective, or eventually SPT and augers (the latter constitutes a good complement to enable an interpolation between cored boreholes).

12.3.3.2. *Setting up investigations boreholes*

The boreholes are drilled according to the results of the geophysical explorations, as they contribute to its interpretation just as geophysics helps extrapolate their results. The desirable number of boreholes depends greatly on the geological complexity of the site and the degree of prior knowledge on it. As a guide, we can consider that at least one point of information is necessary for every 30 to 50 m.

In case a geophysical investigation cannot be performed, at least two points of investigation are necessary, one near each extremity of the work going down 2 to 5 meters under the level of planned trajectory. Boreholes which are not drilled down to the required depth do not provide any specific information in respect to horizontal drillings.

Between the two extremities, other boreholes should be set up along the path in staggered rows at a maximum of 5 meters from the proposed line. The intervals between the boreholes should be decided in such a way as to enable a good appraisal of the possible changes of the ground and to detect the presence of obstacles or cavities, even where the plan crosses a watercourse.

Urban cluttering is often an obstacle to the layout of boreholes; note, however, that it is not always vital to drill boreholes exactly on the planned path:

– first because they have a very remote chance of encountering local heterogeneities;

– second, because information outlining the reality is very often more precious: for example, it enables us to measure the transverse dip of the layers, or sometimes even to improve the horizontal alignment.

Similarly, it is important to extend boreholes several meters below the envisaged maximum depth for the drilling in order to gain a better understanding of the configuration of layers and to make the interpolations between boreholes more reliable.

12.3.3.3. *Test drilling methods*

The following test drilling methods are at the disposal of a geotechnician.

12.3.3.3.1. Test drilling with sampling of the soil

Trial pits

These enable a direct visualization of the soil, its photography and recovery of intact and/or remolded samples from the inner walls. They are useful for small dimension drillings, which are close to the surface and to recover gravel and pebble samples.

Except for shielding devices, they are limited in depth to the first few meters of ground.

Auger drillings by hand

These enable the recovery of remolded samples of fine-grained soils, extend investigations to congested and less accessible parts of trajectories and deepen pit explorations beyond the pit bottom. The investigation depth is limited to a few meters.

Sinking window samplers with percussion hammers

These call for the same observations as for auger boring by hand along with the investigation depth that can reach about 10 meters under certain conditions.

The samples enable a better appreciation of the structures of the soils.

Continuous auger boring

These require the use of a drilling machine, and enable the recovery of remolded samples from fine-grained soils (sand and clay). They provide only an approximate idea of the depths of ground changes and present difficulties in recovery of samples from grounds under the water table. They are largely used for *in situ* tests.

Percussion cored surveys

By hammering, a coring tube or barrel (U100) is driven into the ground with a dropped weight; they are little used in France and tend to destroy the structure of the samples of the recovered grounds. They are not very useful in soils with important inclusions.

By piston sampling and vibratory-drilling, the coring bit (simple, double or triple tubes) is pushed into the ground by a drill equipped with a specific head.

In fine-grained soils without too many inclusions, they enable fast and continuous recovery of core samples and of intact samples.

Surveys by rotary-cored drillings

These need a drilling machine that is equipped with a rotating head, and enable continuous recovery of core samples of soil and intact samples for laboratory tests. By adapting the coring bit, drillings fluids, and the particular type of core-barrel, practically all types of soils can be recovered intact; even sands and gravels (using a double core barrel with split inner barrel or a triple core-barrel). The use of a split core barrel or a triple core barrel is the only means for recovering rocks in their original state and preserving their fissuring for observation.

They are not suitable to correctly sample coarse-grained soils (for example, pebbles and blocks) because of the limited diameters of the coring bits.

12.3.3.3.2. Surveys without soil recovery

Destructive drillings by roto-percussion

These provide a driller's borehole log of the ground. The method consists of interpreting the drilling spoil or cuttings, and requires an experienced drilling team. It should not be used alone without other methods of survey.

It must be considered only as a control drilling and/or for carrying out *in situ* tests.

Destructive drillings by roto-percussion with recording of the drilling parameters

This method is identical to the preceding one, with the same constraints, but provides information on the progress of the drilling enabling a post-analysis of the drilling. The recorded parameters must include: the instantaneous speed of advancement, the weight on the tool, the torque and pressure of the drilling fluid.

Pilot investigation drillings

These are parallel pilot drillings or drillings on the axis of the horizontal drilling with recording of the drilling parameters and recovery of the returning drilling fluids, enabling an investigation on the whole length of the horizontal drilling.

Surveys must be carried out strictly in accordance with national and European regulations and standards.

During the horizontal drilling, to avoid the problems of a rise in the level of water or drilling fluid by the surveys, the investigation boreholes must be backfilled at the end of the investigation work or before the starting of the horizontal drilling (surveys can be suitable for checking the water table level).

12.3.3.4. *Samples for laboratory tests*

Intact and remolded samples of ground must be recovered in order to carry out tests in the laboratory. The tests must include:

– for the fine-grained soils, the determination of:
  - the geotechnical identification,
  - the moisture content,
  - Atterberg limits,
  - the *in situ* density,
  - simple compression strength,
  - particle size distribution, sedimentation;

– for coarse-grained soils, the determination of:
  - the geotechnical identification,
  - the particle size distribution, sedimentation,
  - the grain shapes,
  - the relative density,
  - the abrasive capacity;
– for the rocks, the determination of:
  - the geological description, the stratification and the fissuring,
  - the density,
  - the simple compression strength,
  - the abrasive capacity and crushability.

Samples of water coming from the water table and water intended to be used as drilling fluids will be recovered. Physicochemical tests will include the determination of:

– the conductivity,

– the pH,

– the organic matter content,

– calcium, magnesium and potassium levels,

– sulphate and chloride levels; descriptions and the methods to be followed are given in the National and European standards.

12.3.3.5. In situ *tests*

In situ tests are mainly of two types.

12.3.3.5.1. *In situ* tests to determine the ground parameters which are not easily obtained in the laboratory

Examples of tests of this kind include:

– the determination of *in situ* density of granular soils,

– the determination of the permeability of the ground.

12.3.3.5.2. *In situ* tests which supplement the investigation by boreholes

The results of *in situ* explorations must be compared with those of the neighboring cored surveys and must go down to the same depths.

Useful *in situ* tests with this objective include:

– the *pressuremeter tests* carried out in continuous augur or roto-percussion boreholes especially in granular soils where the determination of the relative density and recovery of samples can sometimes be difficult, as well as in clays and rocks where the pressuremeter characteristics can be compared with the simple compression strength of the materials;

– the *tests with the static and dynamic penetrometers* to confirm the presence of different layers of soils and of the substratum, especially in fine-grained soil. Some dynamic penetrometer equipment can be used to carry out tests in watercourses, investigations that are otherwise too difficult to perform by other methods.

12.4. Contents of the geological-geotechnical dossier of a project

The soils report must import information on the soils in a clear fashion and without any ambiguity on the aspects which affect the carrying out of a horizontal drilling project and within the limits of the means of investigation, which are applied. It is appropriate that the preparation of the investigation campaign and the drafting of the report are entrusted to a bureau of geotechnical studies with specific knowledge in the field of horizontal drillings or to a bureau of study of a specialized firm.

The geological and geotechnical aspects to be specified are:

– the circumstances and the objective of the work to be undertaken,

– a plan of the appropriate scale, on which is indicated the starting point and the end point and possibly any intermediate points and obligatory passages or constraints (existing works),

– a complete description of the surface and of the sub-soil.

The following details are to be looked for initially in existing documents, in particular:

– the geological map of the area,

– previous soil studies in the sector,

– aerial photographs;

and to be complemented by:

– a visual survey of the site and surroundings,

– a survey among the residents around the site,

– a detailed description of the site and of the sub-soil, including its heterogeneity and its stratigraphy. Only a campaign of ground investigation can improve the knowledge on the variations in the sub-soil, their dip, their various horizons and the presence of inclusions, of cavities, of faults, of water tables, etc. These parameters are necessary to define the profile of the drilling,

– the presence of a water table and its chemical composition: for the analysis of the drilling fluid during the drilling of the pilot hole,

– these details, sometimes considered as secondary, are in fact very important during the phases of opening up the drilling and pulling back the pipeline,

– the hardness of the ground (the difficulty to penetrate or to drill the ground): for clays and marls, it can be calculated by the measurement of non-drained cohesion, or the simple compression strength. But the capacity to drill in clay depends also on other criteria, in particular if it is sticky or susceptible to swell, which are related to the clay's mineralogy, its water content and its plasticity (Atterberg limits).

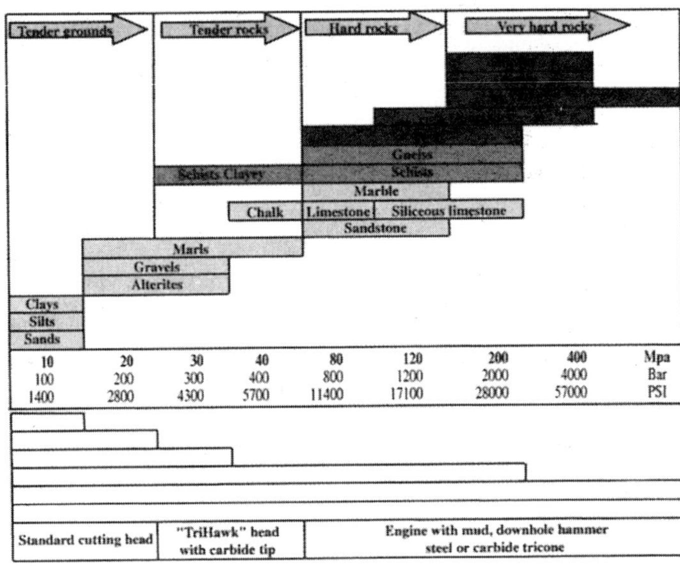

Figure 12.3. *Ground hardness and appropriate drilling assemblies*

The drilling of a granulated soil depends also on other factors, in particular on the lithography, granulometry and the presence of water. In rocks, the measurement of their simple compression strength enables the selection of adequate equipment and permits an estimation of the rate of the progress to be reached.

The *abrasiveness of the ground* is required to estimate the abrasion of the bit and the drill rods.

In addition to the immediate results of the exploration, it is advisable to draw longitudinal sections of the trajectory of the envisaged horizontal drilling on a suitable scale, together with a geological and geotechnical profile.

It is recommended to use the soil classifications and descriptions recognized at the national and European level.

Chapter 13

Guidelines for the Choice of Drilling Rigs and Equipment

13.1. General information

Chapter 10 dealt with the principles of operation for horizontal drilling, along with their main functions, which are:

a) creation of the pilot hole,

b) backreaming of the hole,

c) laying of pipes.

The follow-up and the trajectory corrections vary between various drilling rigs and they hardly are a determining factor of their choice.

The experience of sites shows that some problems encountered, particularly in relation to the ground to be excavated and evacuated, could have been restricted or even avoided with equipment that was better suited for the ground. This obviously assumes that we should have a good prior knowledge of these grounds, their nature (clay, sand, rocks), their condition (humidity, plasticity, compactness), and above all their heterogeneities to which these sites are very often highly sensitive; these geotechnical investigations form part of Chapter 12.

However, even when we know the ground well, the choice of drilling rigs and their equipment is often tricky:

– firstly, because the small dimension of structures makes the digging very sensitive to variations in the nature of the grounds, even at decimetric scale, as the excavation section can vary very rapidly from soil to rock (blocks embedded in a matrix), or even from granular ground to a clayey ground (alternation of sand, silt and clay layers);

– secondly, because it is not possible to make modifications to the machine during the cutting, for example, in order to change the equipment (cutting tools), without taking out the whole string of drill pipes;

– and finally, because in case of a major incident (blocking by an obstacle or following excessive friction caused by swelling clayey ground, for example), the installation of a new parallel drilling is costly and time-consuming compared to the initial project.

Thus, following the example of large tunnel borers which are currently the subject of the technological developments aimed at designing "universal" machine, i.e., the drilling rigs capable of drilling in all types of ground, the problems connected with horizontal drilling systems are comparable or even amplified by the specific nature of the drillings.

The object of this chapter is thus to provide some recommendations on guidelines that need to be followed when making the choice of drilling rigs and their equipment, depending on the current state of knowledge and technology of the drilling rigs suggested by the manufacturers. Of course, they are likely to evolve over time, and therefore provide other answers to the questions raised here.

13.2. Choice of drilling rigs according to their power

The results of the horizontal drilling depend on various factors, particularly on those related to the characteristics of drilling rigs:

– pullback force,

– torque,

– capacity of mud pump.

Table 13.1 gives an idea of the diameters and the distances that one can reach under favorable conditions according to the type of machine. It concerns the highest performances which cannot be combined. The diameter is that of the borehole, i.e. 1.5 times bigger than the pipeline.

The drilling rigs are generally classified according to their highest pullback force. Therefore, the following names which appeared are still applicable:

– mini drilling rigs,

– medium drilling rigs,

– maxi drilling rigs,

– mega drilling rigs.

Pullback force (kN)	Maximum diameter (mm)	Maximum length (m)
30	200	100
70	300	150
120	400	200
150	500	250
200	700	400
400	1,000	600

Table 13.1. *Maximum performances*

Figure 13.1. *Medium drilling machine*

13.2.1. *Mini drilling rigs*

These drilling rigs are mainly used in the field of supply in urban area as well as for laying pipelines or cables. They develop pullback forces up to a maximum of 150 kN, torques in the range of 10 to 15 kN.m and weighing up to 10 t. A good number of these drilling rigs are track-mounted chassis (rubber).

13.2.2. *Medium drilling rigs*

These drilling rigs are often used for the small-scale transport of water or for special types works as in the case of sensitive environmental area for example (Figure 13.1). The produce maximum pullback forces in the range of 150 to 400 kN, and torques in the range of 15 to 30 kN.m and weighing from 10 to 25 t. These drilling rigs are also generally track mounted and hence suited to all types of grounds.

13.2.3. *Maxi drilling rigs*

These drilling rigs are used for large scale pipelines and drillings of greater length. They are often used on transport networks, and can also be used for the transport of water, and the installation of railway lines and important crossroads.

Figure 13.2. *Maxi drilling rigs*

Maximum pullback forces of these drilling rigs are between 400 and 2500 kN, torques varying between 30 and 100 kN.m and their weights vary between 25 and 60 t.

13.2.4. *Mega drilling rigs*

This type of drilling rigs is designed for non-standard lengths of crossings and diameters of boreholes. At most, these drilling rigs can develop pullback forces greater than 2,500kN, a torque higher than 100 kN.m and they can weigh over 60 t.

Table 13.2 sums up various classifications of drilling rigs.

Drilling rigs (types)	Maximum pullback forces (kN)	Maximum torque (kN.m)	Weight in tons
Mini	< 150	10 – 15	< 10
Medium	> 150 and < 400	15 – 30	10 – 25
Maxi	> 400 and < 2,500	30 – 100	25 – 60
Mega	+ 2,500	< 100	> 60

Table 13.2. *Classification of the horizontal drilling rigs*

13.3. Choice of drilling rigs according to their technical characteristics

Horizontal drilling rigs mainly consist of a steel framework on which a mobile carriage is mounted. This carriage transmits the required forces and torque to the drill string. The base situated at one of the ends of the steel framework, will enable the slope of the machine to be modified. This is vital in order to respect the defined rack angle of the bore. The main characteristics of a drilling rig are as follows:

a) chassis,
b) transmission of the forces,
c) power limits.

13.3.1. *Chassis*

The chassis of a drilling rig may be mounted in many ways.

13.3.1.1. *Base*

The simplest method consists of mounting the carriage on a steel framework and equipping it with a base at one of its ends. This base enables the rack angle required for the drilling to be maintained. The advantages of this method are: a simple and strong construction as well as a relatively reduced weight. A disadvantage of this method is the need to use motorized widespread lifting equipment for the operations of unloading and assembly or respectively for dismantling and loading. In some cases, the mobile lifting equipment will need to be of a greater capacity.

13.3.1.2. *Trailer*

A motorized widespread structure is the mounting of steel framework on a trailer chassis. This type of structure makes it possible to gain a relatively greater capacity of manoeuvring on stabilized roads and paths at a relatively lower cost. One of the disadvantages is the restricted capacity on hard grounds. Moreover, most of the drilling rigs of this type have their landing on the front edge, in such a way that after a drilling, the tractor engine should be driven on the opening pit zone to fill in. Otherwise, the drilling machine should be taken away by another device (an hydraulic digger for example) so that the tractor engine can go over a relatively clean ground in order to do hooking.

13.3.1.3. *Track mounted chassis*

A structure in the form of track-mounted chassis was introduced in the meantime for the large and medium size drilling rigs (correct distribution of load). Large and very large drilling rigs are mounted more and more on suitable caterpillars. The advantages of this structure lies in its large capacity on hard grounds. By using a mobile engine for the supply of energy, it is possible, as in the case of a wheeled chassis, to get a machine which can be operated on site quickly. The disadvantage lies, besides its heavy weight, in the relatively higher costs.

13.3.1.4. *Wheeled chassis*

Drilling rigs may also be mounted on a motorized wheeled chassis. In principle, the mobile engine is used also for the supply of energy to the drilling rig. This type of structure enables the installation of the chassis on a trailer. The disadvantage lies in the fact that these structures are heavy and costly. By and large, these types of structures have not succeeded in establishing themselves.

13.3.2. *Transmission of forces*

From the point of view of the mechanical transmission of forces within the framework of drilling, it is possible to distinguish between the following arrangements of the structures.

13.3.2.1. *Chain driven*

The transmission through chain is a simple and reliable system of transmission. With the help of hydraulic motors, as in the case of track-mounted chassis, a chain will be made to move. This chain on its turn is fixed to the framework of the rig. This type of transmission has demonstrated its utilities, particularly for small and medium size drilling rigs.

13.3.2.2. *Rack and pinion*

The heaviest transmission is that by pinion gear and rack. Pinions driven by hydraulic motors are fixed on the mobile carriage. They stick teeth on the rack, which is fixed. With this principle of construction, it is possible to transmit very high forces.

The disadvantages lie in the relatively high weight and a low average speed of the displacement of the carriage.

13.3.2.3. *Hydraulic jacks*

Use of hydraulic jacks for the transmission of forces to the mobile structure is one of the latest advancements in the mechanical construction sector. The use of jacks was already in vogue for the small and medium size drilling rigs. It is becoming more and more common for the large and very large drilling rigs.

Hydraulic jacks may be handled very accurately and in addition to that have a good force/speed ratio. Their relatively high fragility is a drawback, particularly through the cylinder, which is often exposed "in extended position". Repair work can rarely be performed in the site due to the fact that only specialized workshops can repair jacks of such big size.

13.3.3. *Power limits*

Power limits of a drilling machine depend on the mechanical parameters of the machine, but also on the hydraulic capacities of the pumps and the equipment used to carry out the mixtures. In addition to this, the underground plays a very important role in the determination of the lengths and diameters of the possible drilling. It is also necessary to take into account the parameters linked to the pipe such as the mass, roughness of the surface, etc. as well as the plan of the drilling axis, the minimal radius of the curve, etc.

13.4. Drilling rods

These are the essential tools which make it possible to:

– push the drilling head,

– rotate the drilling head and its tools,

– direct the drilling in vertical or/and horizontal curve,

– transport the drilling fluid,

– pull out the backreaming tools,

– install the final product pipe.

Figure 13.3. *Drilling bars*

They are henceforth reliable, as long as it is checked whether they are suitable for the radius of the curve and torque which are needed for the project. They are most frequently imported from various countries of Europe or America. They are of two types:

– cast as a mono block called "forged",

– with pin and box joints welded by friction on a metal rod, often with a steel of different quality.

The treatment of the steel at the weld and its surrounding area around is essential.

Of different length and diameter, each one has its own threading corresponding to each boring tool, with or without patent. The API (American Petroleum Institute) standards specify the minimum admissible characteristics and define some types of threading. The diameter of the joints may be greater than that of the rod, offering planes which make it easier to hold the tools of tightening/dismantling and assemble various attachments such as drill holders, backreamers with or without incorporated swivel, etc.

If any drilling rod "breaks", this almost always of results from the efforts that the driller has thrust on that rod, beyond its limitations of usage prescribed by its manufacturers:

– non-checking of rods after utilization (jacks/cracks/gimlets),

– making up torque,

– work torque,

– limitation of the curve radius.

A curve which is too sharp produces efforts and wear and tear of the bar and its threaded edges, which shall be found only during further drillings. Only the first rods which are at the tip of the drilling, suffer significantly: a check and a rotation in the drill string make it possible to avoid minute cracks and premature loss of the train of rods. The threadings should get only lubricant (copper or zinc), which will guarantee making up and breaking the joints without freezing and/or "blackening". Rare incidents generally occur halfway down the thread of the pin, and/or by turning the box into "the form of tulip". If the quality of steel is not checked by a metallurgical test, it is to the torque (too strong or too weak) and to the drilling work or to the radius of curve which is too sharp that one can attribute the breaking or the deterioration of the rod, and therefore of the whole drill string, in most cases.

13.5. Tools

Depending upon the nature, hardness, abrasiveness of the grounds, various types of tools can be attached to the drilling machine:

– a drill head with blade for the soft ground,

– a drill head with tungsten carbide pick for the hard grounds,

– a mud motor which is hydraulically or mechanically driven and equipped with a drill bit,

– a "down the hole" hammer activated by a compressor for very hard rocks.

There are also a number of types of backreamers, which are described below.

13.5.1. *Wing cutters*

The teeth of wing cutter cut down the materials into shavings and mix them in the drill fluid. The space between the blades and the exterior crown enables the mud formed to get evacuated towards the rear. They are recommended for use with homogeneous materials: clay, compact sand.

Figure 13.4. *Wing cutter*

13.5.2. *Spiral compactor bells*

These are specially adapted for heterogeneous grounds, as their spirals make it possible to evacuate stones and gravel, compressing them away. In cases of big diameter and in the case of sticky grounds, their large contact surface may induce strong friction, and delay progress.

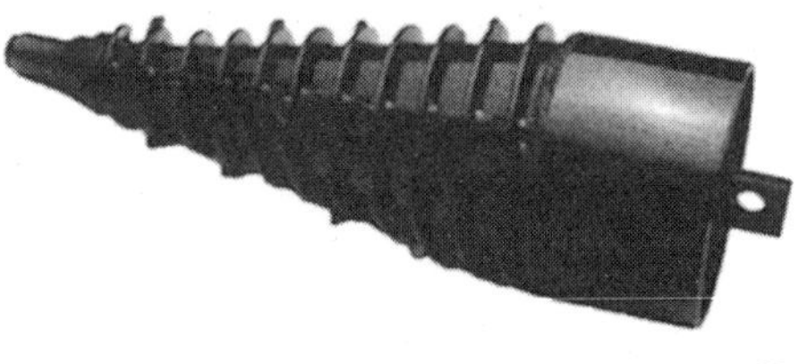

Figure 13.5. *Spiral compactor bells*

13.5.3. *Fluted reamers*

Being an intermediary choice between the two previous models, these are multifunctional. In fact, their helical cone shape enables them to make the ground compact, whereas the picks, numerous nozzles ensure the mixing of mud and its evacuation towards the rear.

Figure 13.6. *Fluted reamer*

13.5.4. *Rock reamers*

These are fitted with rotary cutters made of steel or carbide depending on the hardness of the rock. In the case of abrasive rocks, the body of the reamer and the rotary cutters are protected by the covers of tungsten carbide. Big nozzles ensure the transport of slurry towards the rear.

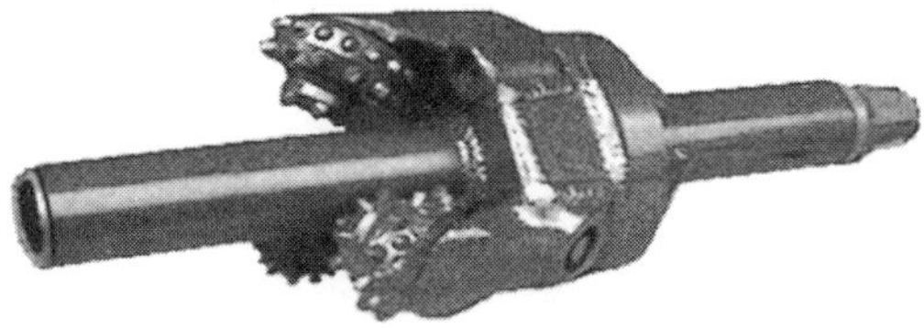

Figure 13.7. *Rock reamer*

13.5.5. *Barrel reamers*

These make it possible to polish the inner sides of the tunnel and are used in the final stage at the time of pulling of pipelines.

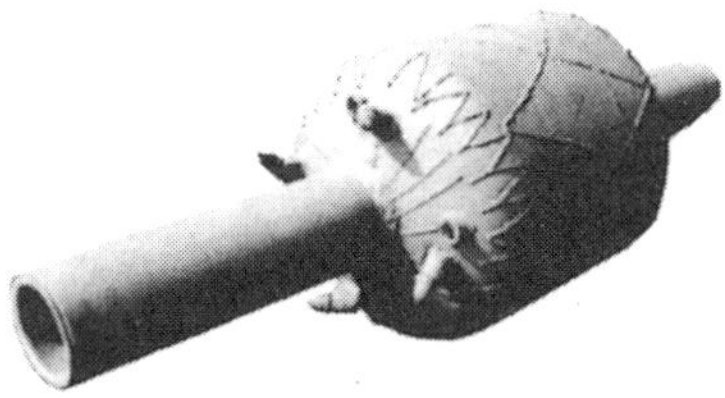

Figure 13.8. *Barrel reamer*

There are also other tools such as rotating joints or swivels and rod recyclers. Swivels enable to pull a pipeline or a drill string without making them rotate. For security reasons, it is preferred that the swivel should be integrated with the reamer. Otherwise, a risk may arise: it may position itself askew and transmit a rotation movement to the rods, which it controls, which may be dangerous to the workers on the site.

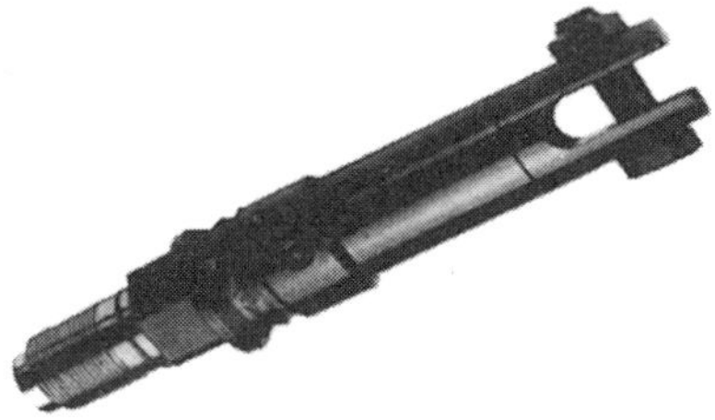

Figure 13.9. *Swivel*

Rod recyclers make it possible to pull a drill string behind a backreamer before backreaming.

Figure 13.10. *Rod recyclers*

Chapter 14

Guidelines for a Project Design

14.1. Basic principles of a pilot pattern

The starting point in the design of the initial pilot hole study is the determination of the drilling axis between the entrance and exit points. It is therefore necessary to comply with some rules in order that refine a drilling axis can be feasible in practice.

It is important to specify at least:

– the rack angle and the exit angle,

– the first and last part of the drilling,

– the radius of curvature,

– the roofing over the microtunnel,

– the ratio between the diameters of the pipeline and diameter of microtunnel after reaming.

14.1.1. *Rack angle and exit angle*

This angle is related to the diameter of the pipeline to be installed, and is generally between 10 and 30% (6 and 15 degrees). The principle is as follows: the more the diameter of the pipeline increases, the smaller the angle will be.

The rack angle (or entrance angle) may be reduced even down to zero, for small drilling machines and for downhole machines.

The exit angle may be increased for pipelines which have a low radius of curvature (PEHD), and therefore have a low (or no) catenary support height.

14.1.2. *First and last part of the drilling*

For the very first segments of drilling, as well as the last, we must avoid imposing a curvature. In fact, the stresses resulting from a high radius of curvature on the rods will modify the profile by increasing the radius, and will be transmitted to the rig itself, due to the low density of the soil.

The length of these first (last) segments varies according to the drilling dimensions, weight, and stiffness of the pipelines to be installed. The longer the drilling and the longer and stiffer the pipeline is, the longer these first and last segments will be.

To give an idea on this, for long drillings, there may be segments without curvature of 10 to 20 m but for short drillings, these lengths can be reduced to 4–5 m.

14.1.3. *Radius of curvature*

To determine the drilling axis, the minimal permissible radius of curvature plays an essential role. It is important to distinguish between the minimum acceptable radius of curvature with regards to the rods (in this case it is the rod manufacturer's allowable minimal acceptable radius) and the minimal permissible radius for the pipes to be installed.

As a general rule, for short drillings or PE pipes, the minimal radius to be taken into consideration in the design is that of the rods (according to the manufacturers, and the type of rod, we may have radii of curvature from 25 to 250 m). For long drillings, as well as steel pipelines, it is necessary to take into account the minimal radius of curvature of the pipeline.

In this last case, where the minimum radius of curvature is related to the pipeline to be installed, this radius will be determined in a more complex way. It will be necessary to combine the drilling radius of curvature and that which is opposite, of the catenary support external to the drilling.

In general, the calculation of the flexibility curvature is done with an analysis in R_{min}:

$$R_{min} = \frac{E*Da}{2*\alpha_{aut}} = R_{elast}$$

with:

– R_{min}: minimum radius of curvature (m),

– E: elasticity modulus (N/mm²) = 2.06 * 10^5 (N/mm²),

– D_a: outer diameter of the pipe (m),

– α_{aut}: permissible bending pressure (N/mm²),

– R_{elast}: flexible radius of curvature (m).

14.1.3.1. *Radius of curvature of the pilot hole*

For a pipe subjected to an inner pressure, and implicitly taking into account:

$$\alpha long = \alpha aut/2$$

To determine the minimum radius of curvature, the equation becomes:

$$R_{MIN} = \frac{E * Da}{\alpha aut}$$

for a welding joint coefficient $V_N = 1$.

We obtain $\alpha_{aut} = K/S$ with:

– K = minimum bending limit (N/mm²),

– S = safety factor.

Thus, the minimum value of the radius of curvature will be defined for the steel pipelines by the following formula:

$$R_{min} = \frac{2060 * S * Da}{K}$$

This formula should only be used for nominal diameters < DN 400.

For larger diameters, it is necessary to rely on the final report of the working group of the company Ruhrgas AG of 1996, and to use the following formulae:

Nominal diameter of the pipe	**Formula**
400 < DN < 700	$R_{min} = 1,400 * D_a^3$
700 < DN < 1200	$R_{min} = 1,250 * D_a^3$

NOTE: the interpretation of these formulae may vary in every country.

14.1.3.2. *Combined radii*

The previous recommendations relate to the absolute minimum radius. If the drilling axis is perfectly linear in the horizontal plane, the minimum permissible radius of curvature is then equal to the minimum radius of curvature in the vertical plane. If we add a curvature in the horizontal plane, the minimum permissible radius of curvature will always be greater than each radius taken separately.

For the study and realization of horizontal drillings, this means the *minimum permissible radii of curvature always correspond to the combined radii of curvature!*

To ensure that the combined radius of curvature of a drilling work is always greater than the minimum radius of curvature, the following formula may be used:

$$R_{comb} = \frac{\sqrt{R_h^2 \times R_v^2}}{\sqrt{R_h^2 + R_v^2}}$$

with:

– R_{comb}: combined radius (m),

– R_h: horizontal radius (m),

– R_v: vertical radius (m).

14.1.4. *Roofing*

The distance between the drilling axis and the ground surface or the streambed is generally called "roofing". We need to note that by convention, "roofing" is the distance separating the surface or the level of the streambed from the upper generatrix of the pipe, or even the top of the borehole.

Experience shows that this roofing should be between 10 and 15 times the diameter of the pipeline. For example, experience shows for a pipeline of a diameter 800 DN, the distance under the surface of the ground or the streambed should be 8 to 12 m.

If, furthermore, for small size pipes, there results a roofing of less than 5 m, it will be necessary to be careful: in fact, such low roofings can cause some raising of the drilling mud, which may be dangerous. Comparable roofing values will need to be kept for the crossing of main traffic roads, landing strips, etc.

National regulations exist for railroad crossings. The more important the roofing, the less dangerous the raising of drilling fluids and sinking of the soils will be.

14.1.5. *Relation between the diameters of the pipeline and the borehole*

The reaming coefficient is the borehole diameter divided by the diameter of the pipe after reaming to be installed.

We must take into account the most important pipeline diameter (if need be at junction points). A correctly chosen reaming coefficient has important consequences for the proper progress during the pipeline-pulling phase. Experience shows that coefficients of about 1.2 (for stable grounds with low frictional coefficients) and of 1.5 (for unstable grounds, the walls of which tend to give way, and for grounds with high frictional coefficient) were convenient.

14.2. Drilling plans

In horizontal drilling, different types of representation are necessary for the project study, such as:

- longitudinal profile,
- plan view,
- cross-sections.

For long drillings, it is also recommended to have:

- the site installation plans,
- the catenary and the launching ramp.

14.2.1. *Longitudinal profile*

The representation of a longitudinal profile of drilling must include at least the following information:

- groundwater,
- topography of the soil,
- surface of the waterway, and level of the different stream beds. In the case of tides, the highest water level and of the lowest water level,
- the entrance angle and exit angle of the drillings,
- the drilling axis, regularly marked (for example, every 10 m),
- the data relating to the vertical radii of curvature for each segment,
- the data relating to the combined radii of each segment,
- the data relating to the horizontal length and relating to the total length of the drilling,

– the roofing indication in the critical zones, such as under streams,

– the position and the depth of the core-sampling and other test drillings, as well as the data on the different layers of soil encountered,

– the known obstacles, for example, network pipelines, massifs de foundation (foundation bed), sheet piles, etc.

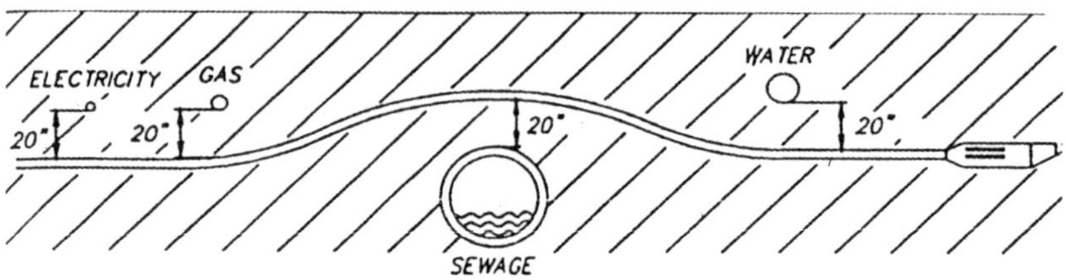

Figure 14.1. *Example of a longitudinal profile with installations nearby*

14.2.2. *Plan view*

The plan view of a horizontal drilling must integrate at least the following information:

– the topography reaching a distance of 5 to 20 m on each side of the drilling axis,

– the recording of approach and exit points of the drilling,

– the drilling axis,

– the data on the horizontal radii of curvature on each segment,

– the positions of core-sampling and other drillings,

– the known obstacles, for example, network pipelines, massifs de foundation, curtains of sheet piles, etc.,

– the work surfaces planned for the drilling machine and the pipeline,

– the orientation in relation to North.

14.2.3. *Cross-sections*

These section representations must include at least:

– the diameter of the horizontal drilling,

– the section of the pipe(s) accompanied by the technical characteristics of the pipe, such as the type of material, the thickness of the pipe, the covering and the mechanical protection, or even its reinforcement.

14.2.4. *Work site installation plans*

The minimum information required will be:

– the position and the importance of the main elements of the drilling machine, including the maxi rigs, the control cab, the energy feed, etc.,

– the type of fixing of the drilling machine,

– the position and the importance of the drilling fluid retention basins,

– the stocking and workshop surfaces.

14.2.5. *Catenary and launching ramp*

The plans for the catenary and launching ramp must contain:

– the catenary radius,

– the position and number of rollers in plan view, or even a top view,

– the maximum height of the upper arch.

14.3. Design notes

In horizontal drilling systems, the study of the project necessitates the testing of numerous calculations. These computations are the responsibility of the contractor or that of the consultant, depending on the phase concerned.

There are two stages:

1) the work stage,
2) the operations stage.

The work phase calculation is generally the responsibility of the performing contracting company, while the operations stage calculation is the responsibility of the consultant and/or the owner.

14.3.1. *Calculation for the work stage*

Installing a pipeline in horizontal drilling systems raises the following questions:

1) will the pipelines to be installed resist the expected solicitations during the pulling phase?

2) is the technical equipment suitable to provide the pulling forces necessary?

To answer these two questions, we need to determine the maximum force expected during the pulling phase. We need to make the difference between the forces necessary at the drilling head to overcome the frictional forces due to the pipeline, and the forces that need to be provided by the drilling machine; in fact, the forces that must be provided by the drilling machine are always greater, for they must overcome the additional frictional forces (rod train, reamer).

14.3.1.1. *Pulling forces at the level of the drilling head*

During the pulling phase, part of the pipeline is positioned inside and the other part on the outside, on the launching ramp or, for short drillings, on the ground.

The forces enabling to overcome the frictions exerted inside the hole depend on:

1) the frictions between the pipeline surface and the drilling fluids: the latter strongly depend on the type of pipeline coating and the drilling fluid parameters (density, viscosity, proportion of the materials transported in the fluids, etc.);

2) the frictions between the pipeline surface and the inner surface of the borehole. These depend on the parameters below:

- sub-soil parameters (frictional coefficient),
- forces resulting from the weight and buoyancy inside the hole,
- geometry of the whole borehole (radii of curvature).

The forces enabling to overcome the frictions at the launching ramp depend on:

– the pipeline weight with its coating, the mechanical protection, and the buoyancy system,

– the type and geometry of the bearing rollers,

– the catenary radius,

– the length of the pipeline lying on the launching ramp,

– the physical conditions of the launching ramp.

14.3.1.2. *Tractive forces at the level of the drilling machine*

The drilling machine provides pulling forces that are greater than the forces encountered at the drilling head. This is due to the fact that all the parts of the rod train are subjected to frictional forces.

The value of these frictional forces partly depends on the mechanical parameters associated with the drilling machine, such as the specifications at the level of *the tool joints and the weight of the rod train.*

This value also depends on the borehole by its *geometry*, on the *radius of curvature* (the more important this is, the less important the frictional forces are), and finally on the *pulling phase* (the frictional forces at the head decrease with the penetration).

14.3.1.3. *Calculation methods of pulling forces*

The two methods used most frequently to calculate the tractive forces during the pulling phase are:

– the method complying with the Netherlands standard NEN 3651,

– the American method AGA (*American Gas Association*).

With these two methods, it is possible to determine the maximum forces that are likely to apply at the drilling head.

Empirically, we may say that this maximum value is reached right before the end of the pulling phase, when almost all the pipeline is inside the hole.

14.3.1.4. *Calculation of the drilling machine dimensions*

There is not yet a universal calculation method to determine the dimensions of the drilling machines. If this were the case, the method should result, during the pulling phase, due to the instability of the hole, and sometimes due to the prolonged interruptions, in a pulling force considerably greater than that which is determined by the theoretical calculations.

As a reference value, it is recommended to retain, according to the nature of the formation, a safety coefficient of between 2 and 3 between the theoretical force and the capacity of the drilling machine.

That means, for calculated pulling forces of 300 kN, we must use a drilling machine with a capacity of 600 to 900 kN.

14.3.1.5. *Supports*

Adapted supports must be able to take up the tractive forces of the drilling machine, forces that are sometimes very high.

It is recommended to calculate the dimensions of these supports according to the maximum tractive forces applicable to the drilling machine.

14.3.1.6. *Stresses suffered by the tubes*

The maximum tractive forces are also used to determine the stresses exerted on the pipes. The longitudinal stresses must be taken into account, as well as the stresses from the bending (stresses according to the drilling radius of curvature).

14.3.1.7. *Protection against collapse*

For the PE pipes in particular, it is usually necessary to show the proof of the integrity of the pipeline to be installed (outer bending and pressures).

It is possible to compensate the outer pressure of the drilling fluids with an open pulling head enabling fluids to enter the pipeline.

Another possibility is to fill the pipeline with water.

During the construction phase, it is possible to use the short-term elasticity modulus of the PE, a value given by the pipeline manufacturers.

14.3.2. *Calculation of operations stage*

During the operations stage, the maximum pressures are applied inside the pipes. The stresses *resulting from the bending* (that is, the drilling curvature) will essentially be taken into account.

The stresses related to the inner pressure replace the pulling stresses that have become non-existent.

The proof of safety against the geometrical deformations must take into account the elasticity module in the long term (PE pipe).

From time to time, it is necessary to verify the lifespan of the pipeline inside the hole in different conditions. This is even more important where the environment is particularly corrosive, or very hot.

14.4. Work planning

The work planning will contain at least the following stages:

– site installation,

– pilot hole drilling,

– backreaming,

– pulling,

– pipeline construction, or welding,
– site installation disengagement,
– surface mending.

Representation of the time schedule in linear form is generally suitable. The time unit will be by working day.

With such a representation, it is easy to make comparisons between the penetration planned and achieved, and to draw the consequences of possible delays in the work.

In addition to these "relative" data, planning must also take into account more absolute data:

– the "earliest" date of the beginning of work,
– the "latest" date of completion of work.

14.5. Drilling fluids

14.5.1. *General information*

Resulting from the petroleum technique, the drilling fluids, commonly known as "sludge", slurry or "mud", are complex fluids that play a vital role in the construction of several sites, particularly trenchless work projects, whose success is largely influenced by them.

We can find a detailed description of the different functions and properties of drilling mud used for large sludge pressure tunnel shields in the AFTES Recommendations (2002).

We recall that, generally, the fluids used for boring may have several essential functions:

– maintaining the cutting in suspension and ensuring its removal by hydraulic channels: this obviously is a function that is directly applicable to boring machines with hydraulic mucking;

– guaranteeing the stability of the bore, strengthening the walls and preventing loss of fluids by creating an external or internal "cake" that is as fine and as resistant as possible. This is a supporting function;

– lubricating and cooling the tools, drilling strings, on-board equipment and pipelines;

– facilitating digging by jetting. This function is sometimes necessary in clayey ground (see Chapter 5).

There currently exist many products in the market, and the choice of a bore fluid formulation suitable for a given project is based on experience.

Air-based fluids such as foams are excluded from this discussion because they are rarely used in drilling techniques. They are however the subject of research as part of the National Microtunnels Project, whose results are described in the Quebaud thesis (1996) and in the ESIP document of Anne Pantet.

The recommendations of the AFTES (2002) highlight the interest of "sludge programme", whose objective is to optimize the choice of the type of sludge and its monitoring and inspection, in order to respond as well as possible to the technical and economic requirements of the project. We therefore need to gradually draft a summary document on the formulation of sludge, its interaction with the natural environment, its implementation, management and treatment.

We must bear in mind that, during the progress of the site, the properties of the sludge change: they obviously depend on the initial composition defined during manufacture, but also on the water and ground debris that progressively add up to the sludge: there is a need therefore to be concerned by the fluid at the start called "clean sludge", but also with fluids polluted with excavation debris "polluted sludge".

The following are the main characteristic parameters of drilling mud, which determine its behavior and which must be regularly measured and recorded as the digging work progresses:

– the density, which is an index of the content of solid element in the polluted sludge; it must generally be between 1.0 and 1.2;

– the viscosity, which characterizes the ability of forming a cake as well as the ease in transportation of the mucking; measured at the Marsch cone, it must generally be between 32 and 40 seconds in clayey ground, and greater than 50 seconds in sandy ground;

– the yield point, the thixotropy and the filtrate that determine the formation of the cake and its ability to reform rather rapidly; in a filtration test, clean sludge must present a cake less than 4 mm and a filtrate less than 40 cm^3; in polluted sludge the cake must remain less than 3 mm, and the filtrate must be in the region of 6 cm^3 in clayey ground, and 10 to 15 cm^3 in sandy ground;

– the sand content, which results from the separation result of solid earth and which affects the permeability of the cake and therefore its stability; it must generally remain less than 4 to 5 % (measured with the elutriator);

– the pH, which affects the ionic balance and thus the physico-chemical properties of the sludge; it must remain within a range of 8 to 10;

– the conductivity and the hardness are also indices that may be useful.

NOTE: the values indicated above are only to indicate the order of magnitude generally used in trenchless work. They must however be adjusted according to predominant performances, which are dependent on the ground. We will give some indications on the expected performances in paragraph 14.5.2 below.

We can also modify during the construction at the site the properties of the polluted sludge, in order to restore the desired properties which may have progressively deteriorated: depending on the rheological and filtration properties required, a simple suitable physical or chemical treatment may be carried out.

Finally, the residual sludge, i.e. sludge that cannot be reused for the site, will have to be treated or eliminated; this operation has become very restrictive due to the change in legislation. It will be dealt with at the end of this chapter.

14.5.2. *Selection criteria*

The level of characteristics required will depend on the difficulties of the project: geometry, complex and varied geology, unfavorable geotechnical elements, abnormalities, pollution of ground.

The greater the number of difficulties, the more precisely the minimum required criteria will have to be specified and analyzed before and during the work.

A priori, the selection of the type of sludge is done by answering the following main questions:

– will the sludge be recycled or not? This is often dependent on the estimated sludge volume of the equipment used;

– what are the main functions desired: stability, mucking, lubrication, ease of removal, etc.?;

– what are the quality criteria for the composition of the sludge?;

– what is the degree of complexity of the implementation and checks to be carried out at the site?

In practice, this selection is often dependent on:

– the nature of the project and levels and quality of information available (study of the ground, etc.);

– the equipment, experience and the skill of each contracting company.

Table 14.1, (FSTT, RT 30), extracted from the technical report no 30 of the National Microtunnel Project, specifies the expected impacts of different classes of ground (classified according to NF P 11-300 standard "Classification of materials that can be used in the construction of backfills and layers in road foundations"), and therefore the main functions required from the mud.

Class of ground	Standardised classification of grounds	Impacts on the microtunneling project
A	Fine soil: clayey silt sticking	Tamping with possible jamming
B	Sandy and gravely soil with fines: sand with gravel more or less clayey	Abrasiveness, deviations, and possible implosion followed by jamming
C	Soil including fines and large elements: clay or flint chalk, grinding grit, talus scree, moraines	Abrasiveness, deviations, and possible implosion followed by jamming
D	Soil insensitive to water: sand with clean gravel	Partial to total loss of mud Instability of the walls
R	Rocks: carbonated, clayey, siliceous, saline, igneous and metamorphic	Rapid wear of tools and shafts Contamination by salts (evaporates) Difficulties in directing the drilling in soil-rock interfaces Loss of mud and instability in fractured areas
Special materials	Organic soil, industrial sub-products	Mud loss, contamination by organic matter

Table 14.1. *Impact on the projects according to the expected grounds*

14.5.3. *Products used*

The drilling mud is essentially made up of a stable colloidal suspension in a dispersing agent: water. Two families of colloids are mainly used:

– *mineral*: mainly bentonites,

– *organic*: mainly water-soluble polymers.

This suspension is rapidly altered by solids in the ground and possibly by water contained in the ground to be crossed and the minerals contained in it.

The bentonites are industrial clay of the smectite group. They are characterized by a foliated structure, which is negatively charged on surfaces and positively on fractures. Upon contact with water, the flakes disperse, swell and possibly exchange the charge compensating cations.

Beyond a certain concentration (relatively low) of the order of 4 to 6%, and depending on the quality of bentonite, a stable structure develops and has certain rigidity under shearing.

The bentonites can be combined with additives for various functions: viscosifying, fluid-loss additive, water reducer, clay encapsulator and stabilizer, lubricant.

The most common are water-soluble polymers which, in addition to their ability in increasing the viscosity, present special physico-chemical properties. There exist several types that are natural, artificial or synthetic, which can remedy specific problems relating to certain soil materials, such as:

– sticky or swelling clay,
– improvement in the stability in sand and gravel,
– better resistance to physical or chemical contaminations,
– abrasiveness,
– etc.

The family of products generally used in the market for trenchless work are:

Bentonites	Polymers	Others
Bentonites with high efficiency	Viscosifying polymer of type PHPA or cellulosic	Surface-active lubricant or detergent (care should be taken in the care of PE)

In practice, companies that have not defined their own standards for boring fluids may refer to the NF EN ISO 13500 standard "Boring fluids: specifications and tests" dated September 1998.

Manufacturers provide a certain number of documents that enable the use of the products under trade practices: technical and safety data sheet (NF ISO 11014-1 standards safety data sheets for chemical products dated November 1994), technical notes, application sheets, etc.

14.5.4. *Recycling and processing*

Depending on the volume of the sludge to be processed, a site may decide to work only with clean sludge during the boring stage, or to provide for a treatment and recycling station for boring fluids.

In the latter case, the important steps are: grit removal, desilting and hydrocycloning.

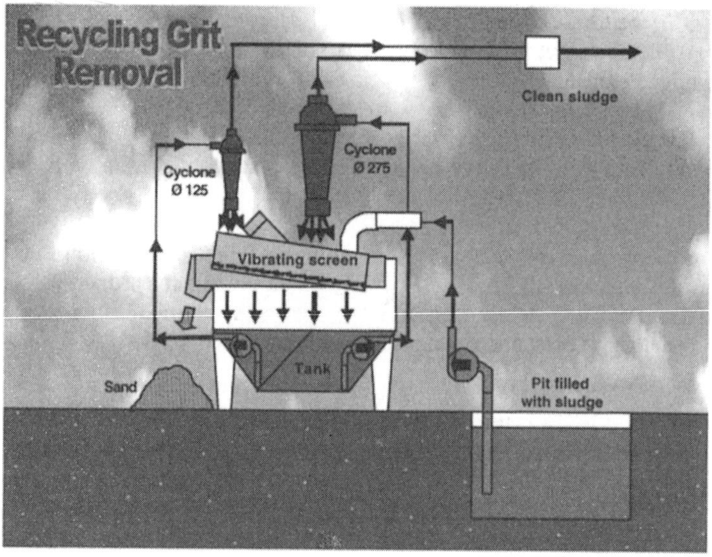

Figure 14.2. *Principle of recycling*

In practice, the earth is separated with the help of vibrating screens, which separate the sand then the silt from the fines, then the cyclones that remove the finest elements by centrifugation. We get a coarse fraction, which can be reused, and a "pulp" made up of the finest elements.

However, the active clay content (from bentonite) must be regularly monitored, as the sludge gets mixed with "inert" fines from the excavated ground, which are not extracted by hydrocycloning and alter the properties of the sludge.

Another solution consists in the addition of a centrifuge enabling the separation of the excavated material from the fluid. The ultimate stage may be attained by installing a "clear water" unit extracting from the water all the particles and thus enabling its direct discharge in the natural environment.

The recycled sludge must therefore be regularly reclaimed, because of this gradual loading of inert fines, but also by contaminations caused by groundwater, and the consumption of bentonite and water by filtration in the ground.

Different types of treatments can help soften the recycled sludge:

– adding new sludge before sending it back into the drilling system,

– adding additives (viscofiers, plasticisers, etc.) to correct the characteristics that have become non-compliant.

14.5.5. *Implementation at the site*

Implementation of drilling mud requires suitable manufacturing, storage and solid treatment equipment:

– mixers,

– main and auxiliary pumps,

– mud tanks, vibrating screens, hydrocyclones, centrifuges, and possibly a plant for the physico-chemical treatment of waste.

The equipment as well as the quality of process water and the temperature will significantly affect the performances of the sludge.

On the other hand, we must also emphasise that the storage conditions (long periods, humid atmosphere, etc.) can significantly alter the characteristics of the bentonite powder.

14.5.6. *Sludge treatment: technical and regulatory aspects*

14.5.6.1. *General considerations*

The purpose of this chapter is to define technical and regulatory criteria for the management of drilling waste according to the regulations in force, and to present the possible lines of treatment for drilling mud.

It includes the main conclusions of the National Microtunnels Project no. 27.

In fact, sludge from treatment plants that cannot be used in that condition anymore, is compared to waste that must be treated in order to be stored. This waste is, either inert or dangerous depending on the cases considered.

Nevertheless, like all waste, sludge belongs to the manufacturer (according to law no 75-633 dated 15th July 1975, modified by law no 88-1261 dated 30th December 1998, law no 92-646 dated 13th July 1992 and law no 95-101 dated 2nd February 1995,), and therefore to the boring contracting company and it must be treated before finally being suitably stockpiled.

Not too long ago, the largely polluted effluents were discharged into the natural environment without any special precautions.

If the purification capacity of the environment (very often the water courses, or waste at the site) was limited, malfunctioning occurred.

This local pollution was partly controlled and the development of pollution removal techniques transferred this essentially liquid form of pollution towards the solid waste sector

Currently, drilling stations are equipped with plants where the fluids are essentially dewatered before disposal on land of various sub-products, or storage units where the residual fluids are provisionally stored before being collected by trucks and taken for disposal, or treatment plants.

Removing water, which is the main enemy of disposal due to the generation of leachates, remains the prime objective.

Waste storers that do not themselves perform the operations of elimination or reclamation must call upon a private or public collecting body.

In case the waste is abandoned, the authority concerned may undertake this elimination at the expense of the responsible contractor/consultant.

In addition, the contractor/mud producer is interested in constituting a reference file, in order to ensure the traceability of products, which must include the information necessary for the identification of the waste produced, and some overall analysis results on representative samples.

However, a drilling site with its treatment tool (treatment and recycling unit) distinguishes itself from other treatment plants by its provisional and mobile nature.

Considering the nature of the sites and the diversity of products, our analysis has led us to define three criteria (see Figure 14.3) to be considered by the company to manage wastes generated by excavation work and to define a suitable treatment line:

– *site criterion*: in addition to the geological conditions: polluted site identified, polluted site not identified, site known as non-polluted,

– *quantity criterion*: site duration, volume to be treatment, treatment location. The volume of sludge to be treated varies from a few m^3 to several dozen m^3 per day,

– *sludge criterion*: mineral sludge, organic polymer sludge, mixed sludge.

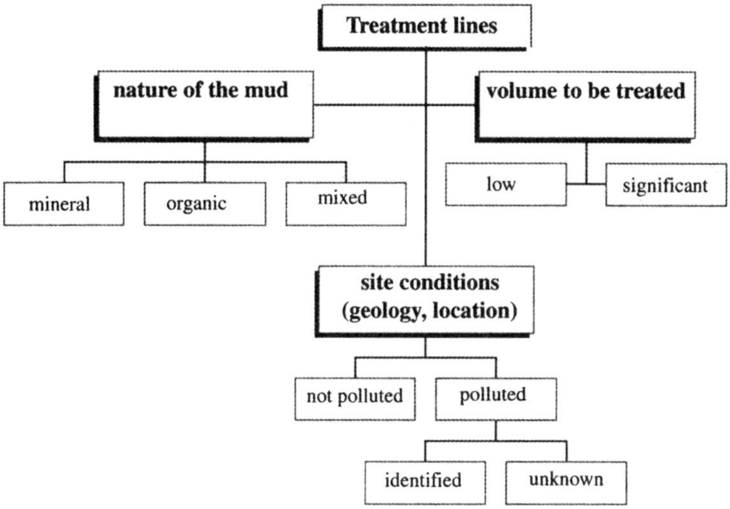

Figure 14.3. *Definition criteria for treatment lines.*

Figure 14.4 summaries the successive stages of the treatment chain.

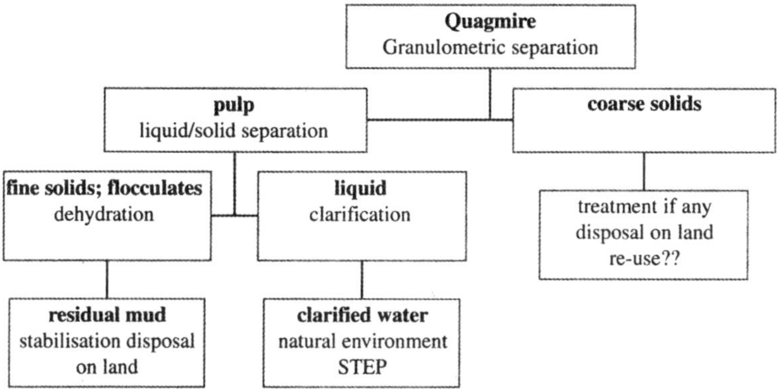

Figure 14.4. *Logic diagram of the main stages of treatment*

14.5.6.2. *Drilling wastes eliminations solutions*

Depending on the waste legislation, the drilling mud and other waste comparable to drilling mud will have to be classified in relation to the source of the bore fluids used and the characteristics of the ground crossed (hydrocarbon, heavy metals and organic matter content). *The soil criterion is fundamental here*, as inert waste mixed with dangerous waste becomes dangerous.

In spite of their essentially inert nature, particularly if the excavated site is identified as non-polluted, the bore fluids cannot be discharged into the natural environment, because their abandonment causes damage to the site or pollution to watercourses by large amounts of sediment loads, which can have a harmful effect on the soil, the flora and the fauna. Several solutions or destinations for the discharge of waste mud, containing soil material, are possible:

– removal without prior treatment: to treatment plant (STEP) or as discharge,

– treatment at the site with the development of a treatment tool at every plant,

– recourse to a collection system within the company or sub-contracted to a specialised treatment centre.

The last two solutions use the same processing techniques (granulometric separation and liquid/solid separation). They are selected according to a criterion for localising the site urban or remote; and a criterion depending on the quantity of mud produced during excavation work.

14.5.6.2.1. Discharge without treatment

Direct disposal into the public water system around the treatment plant of the concerned town may be possible if the waste quantities are small. The sewer systems form part of sanitation installations of communities. They discharge various urban effluents thrown out discharged around treatment plants by individual or some commercial craft and industrial activities. The inflow of such mud may disturb the complex chain for the elimination of pollution.

This solution is discussed on a case by case basis, generally with the local authorities, but after several consultations and considering the composition of mud (too much mineral matter, no organic matter and some metals), we have been advised against employing it by the municipalities treatment plants operators as this waste may cause risks to the proper operation of their plant.

Direct disposal on land without treatment is sometimes used; however, this solution is far from being satisfactory.

14.5.6.2.2. Treatment lines

Generally, the companies must therefore either carry out treatment at the site (liquid - solid separation unit), or send the waste to a temporary storage centre where the mud from various sites is collected, or send it to a processing plant adapted to handle large volumes (this process is particularly profitable in urban areas).

Moreover, if the company decides to develop its own treatment tool, we must then identify whether the site belongs or not to the nomenclature of designated plants.

If it is listed, it must be subject either to notification or authorization and comply with the French governmental regulation dated 2nd February 1998. On the other hand, if it is not so, it must then be subject to the decree dated 29th March 1993 of the Waters Act.

Solid parts (coarse and fine) of the mud

Currently, the drillers negotiate the storage condition of solids. Disposal on land remains the most viable and the most commonly used methods. To be disposed on land, it is necessary that the solids have a dryness of at least 30%. Depending on its nature, the waste could be disposed on waste storages classified as 1, 2 or 3:

Class 3 – (as a general rule): the waste under this class is produced by microtunneling and horizontal drilling operations in non-polluted soil, identified or verified, with the help of bore fluid: water, water/bentonite mixture, without the addition of additives in large quantities.

Class 2 – (occasionally): concerning the drilling mud, this category includes mud that is mixed with large quantities of polymer additives. The theoretical evaluation of the DOC of a suspension of 0.1 % of CMC (1 g/l) gives a value of 740 mg/l. This mud may be considered as being equivalent to fermentive and rapidly changing waste of the industry. This aspect remains to be dealt with in depth.

Class 1 – (exceptionally): the drilling mud coming under this class is:

– drilling mud with hydrocarbon content of more than 1%,

– boring residues resulting from the use of boring fluids with low hydrocarbon content,

– residues from the treatment of polluted soil.

Some polluted soils may be classified as iner, if the results of the polluting potential tests, which include three successive lixiviations according to the standard NF X 31-210, do not show any release of pollutants in the leachate.

Effluents or clarified water

This water can be discharged either directly in the natural environment if the discharge standards and the receiving body of water permit it, or in the public water system towards the municipality's water treatment plant (provided that the standards are respected), or be regarded as a leachate and make them respect the standard in force.

Storage in reservoirs or pits must comply with the legislative provisions in force. It is the same for decanting systems and transportation to treatment plants.

It must be recalled that in the case of water sampling to prepare drilling fluids (either from an unconfined groundwater, or from a watercourse), the company must restrict its water consumption and equip the sampling stations with measurement devices.

14.5.6.3. *Development prospects*

Reuse or reclamation can be envisaged only for waste generated during excavation work done with mineral sludge containing very few additives, in soil having no or very little polluting potential and if the operation is economically viable and technically possible.

If all these conditions are satisfied, the earth of the drilling mud can be reclaimed at the production site itself or after transfer to a platform equipped to gather and process all the waste from a production, equivalent in size to a town.

Some possible methods for reclamation in the field of civil engineering may be envisaged:

– soil mortar ready for use,

– clean material ready to be used for pavement structures,

– granulometric correction of soils, particularly mud containing bentonite could help make some soils watertight if this function is sought,

– use of ultra-fine particles in the concrete.

These techniques are still not well developed, and do not have any immediate application that can be envisaged for waste from microtunneling or horizontal drilling sites.

Chapter 15

Guidelines for the Management of the Site

15.1. Guidelines on lubrication, drilling fluids

15.1.1. *General information*

The drilling fluids fulfill many essential functions:

– maintaining the cuttings in suspension and ensuring their discharge through hydraulic channels, thus cleaning the hole;

– guaranteeing the stability of the borehole, strengthening the walls and avoiding fluid losses by creating a fine and resistant internal or external "cake", thus constructing a ground support by avoiding subsidence and narrowing of the hole as well as seepages;

– lubricating and cooling down the tools, drill string, transmitting probe and pipeline, thereby reducing friction;

– facilitating drilling by the "jetting" effect.

The main characteristic parameters of the drilling mud are:

– density,

– viscosity, stability threshold, rest hardening, etc.,

– filter loss and cake,

– sand content,

– pH,

– hardness.

These different parameters must be measured and recorded regularly during the drilling. Depending on the rheological properties and the required filtration, a simple and appropriate physical and chemical treatment will be carried out.

15.1.2. *Selection criteria*

The greater the number of difficulties, the more the minimum parameters of the drilling mud required will have to be precisely described and analyzed during the work.

15.1.3. *Products used*

In practice, contractors who have not defined their own standards for product requirements used in drilling fluids can refer to the NF EN ISO 13500 standards - specifications and tests carried out in September 1998 on drilling fluids.

The manufacturers supply a certain number of documents that enable the use of their products in good practices: data sheets and safety data (NF ISO 11014-1 standards, safety data sheets for chemical products of November 1994), technical data sheets, implementation sheets, etc.

15.1.4. *Implementation at the site*

The implementation of drilling mud requires machinery, storage and suitable treatment of solids: mixers, main and auxiliary pumps, mud pits, vibrating screens, hydrocyclones, centrifuges, possibly a physico-chemical treatment plant for discharges, etc.

The plant as well as the quality of water processing and temperature will significantly affect the performance of the mud.

The storage conditions (long durations, humid conditions, etc.) alter the characteristics of the bentonite powder.

15.1.5. *Polluted sites, environment, slurry*

During the entire useful life of the drilling fluid, from its mixing to its elimination, it is necessary to assess its impact on the environment according to the criteria listed in Table 15.1.

Regulatory criteria	Technical criteria	Economic criteria
Law on water Law on discharges	Geometry of the site Nature of the ground to be crossed Identified pollutants Type of mud used	Treatment line selected

Table 15.1. *Impact criteria of drilling fluid on the environment*

15.2. Recommendations on reaming

15.2.1. *Reaming diameter*

The reaming diameter (final boring in the case of successive reamings) depends on:
- the diameter of the pipeline,
- the drilling length,
- the nature of the ground,
- the type of pipeline,
- the radius of curvature.

The ratios given below make up a good technical database:
- length less than 50 meters: 1.2 times the diameter of the pipeline,
- length from 50 to 100 meters: 1.3 times the diameter,
- length from 100 to 300 meters: 1.4 times the diameter,
- length greater than 300 meters: 1.5 times the diameter,
- drilling in rocks: 1.5 times the diameter,
- installation of a steel pipeline: 1.5 times the diameter.

15.2.2. *Choice of the reamer*

The type of reamer will depend on the ground. There are in fact three ways to construct a tunnel:
- by cutting the soil, then mixing it in mud to discharge it out of the borehole,
- by compacting the soil,
- by cutting and compacting the soil.

15.2.3. *Multiple bores*

The reaming is only done directly to the final diameter of the borehole very rarely. The reaming is generally carried out in many progressive sequences and it is the train of rods that are pulled during the pre-reaming.

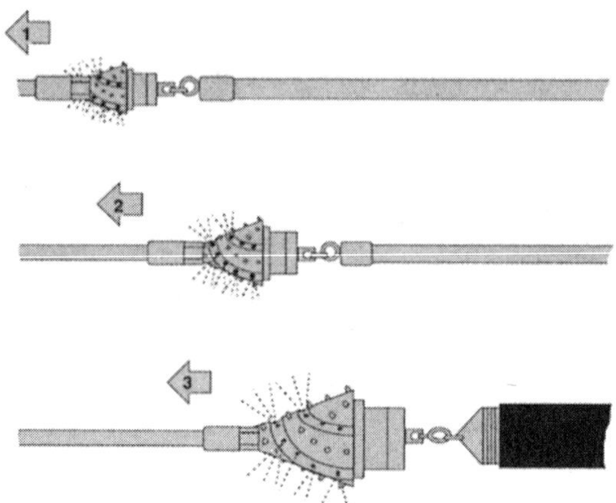

Figure 15.1. *Successive reamings*

15.2.4. *Reaming sequences*

The resistance encountered by a backreamer is proportional to the work area.

The torque of the drill being constant, it is desirable that each backreamer encounters approximately the same resistance.

In addition, to facilitate the discharge of cuttings by mud, its speed must be as high as possible and the volume to be filled and cleaned should be of the same magnitude.

Here too, it is the work area and not the diameters that have to be taken into account.

The area on which the backreamer is operating = the boring area – the previous boring area, which is:

$$S = \pi \times R^2 - \pi \times r^2$$

EXAMPLE: installation of a pipeline of Ø 400 mm. Backreamers available are of Ø 220, 325, 375, 475 and 580 mm. Which backreamer should be selected and which backreaming sequence should be adopted?

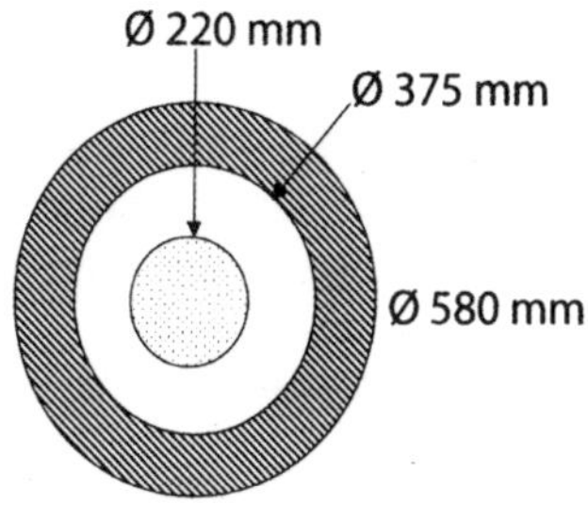

Figure 15.2. *Sequence no. 1*

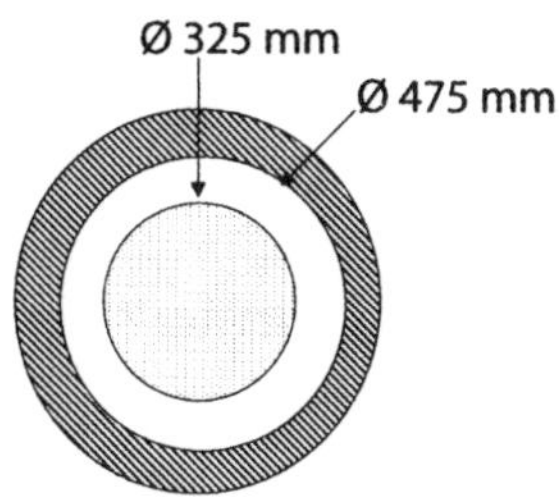

Figure 15.3. *Sequence no. 2*

With sequence no. 1, we get:

– 1st backreaming: r = 110 mm = 1.1 dm; the area is of 3.8 dm^2,

– 2nd backreaming: r = 187 mm = 1.9 dm; the area is of (11 – 3.8) = 7.2 dm^2 with a ratio of # 2,

– 3rd backreaming: r = 290 mm = 2.9 dm; the area is of (26.4 – 11) = 15.4 dm^2 with a ratio of # 4.

With sequence no. 2, we get:

– 1st backreaming: r = 162 mm = 1.6 dm; the area is of 8.3 dm^2,

– 2nd backreaming: r = 237 mm = 2.4 dm; the area is of (17.7 – 8.3) = 9.4 dm^2 with a ratio of # 1,

– 3rd backreaming: r = 290 mm = 2.9 dm; the area is of (26.4 – 17.7) = 8.7 dm^2 with a ratio of # 1.

Sequence no. 2 is preferable because it helps obtain approximately identical areas, thus creating balanced resisting torques for each backreaming. Sequence no. 1 would have imposed a significant amount of stress during the 3rd backreaming.

15.2.5. *Reaming speed*

The boring speed should be compatible with the capacity of the sludge pump. In other terms, the volume of slurry injected in the tunnel must be sufficient to fill up the area dug and discharge the cuttings to the outside. If we consider a "slurry factor" of 3 (the amount of slurry volume required to discharge 1 volume of soil), which is the case in soil with low compressibility soil, the:

volume of slurry injected = 3 × volume dug

or:

volume of slurry injected = rate of flow of the sludge pump × time and volume dug = diameter × length

thus:

rate of flow × time = 3 × boring diameter × boring length

that is:

rate of flow = 3 × boring diameter × boring length/time

or:

rate of flow = 3 × boring diameter × boring speed

The reaming speed must therefore be equal to: boring speed = rate of flow/3 × reaming diameter.

If we consider the third reaming of sequence 2 mentioned above (8.7 dm^2) as an example and if we have a slurry pump of 145 l/min, we get:

reaming speed = 145/3 × 8.7 = 145/26.1 = 5.5 dm/min. = 0.55 m/min.

If the length of the rods is 3 meters, the reaming speed must not exceed 3/0.55 = 5 minutes per rod.

If the reaming is done at a greater speed, there is a risk of the tunnel collapsing.

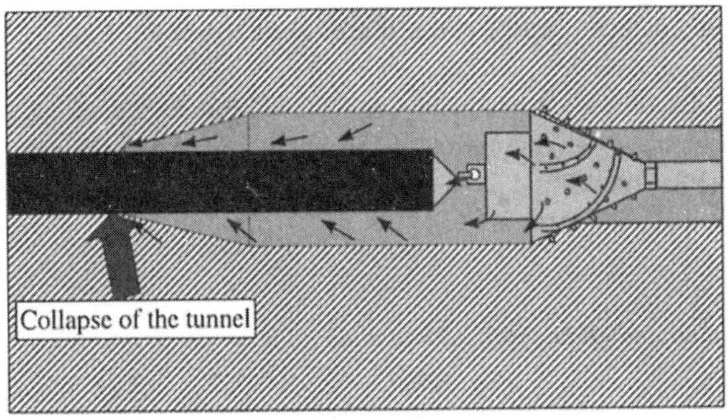

Figure 15.4. *High speed*

15.2.6. *Installing a protective sleeve*

In stony ground, stones can get in between the reamer and pulling head, making the pulling of the pipeline difficult. A solution is to either weld or screw a steel pipe sleeve to the back of the reamer.

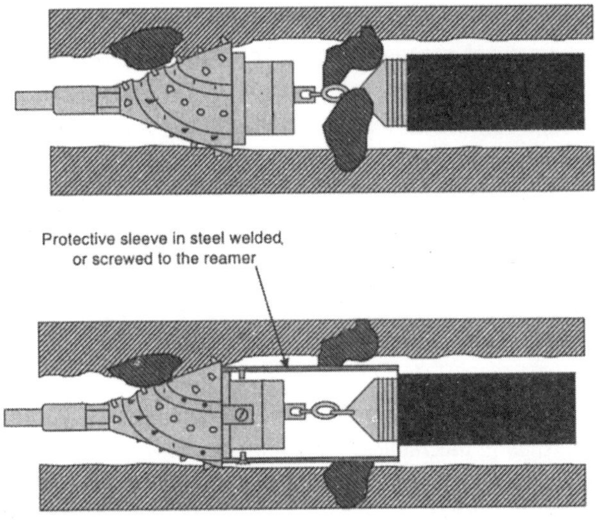

Figures 15.5. *Protective sleeve in relation to stone blocks*

15.3. Guidelines on safety and protection of environment

The constraints linked to the safety and protection of the environment must be considered in the study. All national and European rules must be followed during the work. The elements mentioned below Figure only as additional elements and can on no account substitute the current rules.

15.3.1. *Safety at the work station (at the site)*

Before starting the work, it is necessary for all personnel present at the site to attend an information meeting on accident and their prevention, means of rescue and emergency organizations.

Taking adequate steps may prevent the following hazards linked to horizontal drilling:

– work on inclines,

– work on rotating mechanical parts or tools,

– risk of slipping increased by the presence of drilling mud,

– respiratory risks linked to the inhalation of bentonite powder,

– handling of loads during lifting (drilling rods, reamers, etc.),

– significant torsional moments during the tightening or loosening of drilling rod/tool unions,

– communication between the control cab, the drilling machine and the pipeline side,

– work under thoroughfares,

– risk of stress on underground structures.

15.3.1.1. *Work on inclines*

The working area used on the machines must be of non-skid material and be easy to clean. Fixed handrails must prevent the risk of falls.

15.3.1.2. *Work on rotating mechanical parts and tools*

Contact with rotating mechanical parts must be prevented in every possible way by fixed safety installations. The working clothes of the personnel operating the machines must be tight fitting and closed. During the rotation of the drilling rods, a clearance distance must be maintained.

15.3.1.3. *Risk of slipping increased by the presence of drilling mud*

During the dismantling of the drilling rods, the bentonite mixtures must be collected in salvage tanks. Clean water must be available for the cleaning of the work area near the machines and mixer.

15.3.1.4. *Respiratory risks related to the inhalation of bentonite powder*

Suitable techniques to avoid working as much as possible in areas subject to contamination by bentonite powder must be employed. Work in these areas must be done only by those wearing anti-dust masks (half-masks that filter particles) and airtight protective eyewear.

15.3.1.5. *Handling of loads during lifting (drilling rod, reamers, etc.)*

The handling of drilling rods and other loads by lifting appliances must be done with great care. The proper condition of pipe tongs, slings and straps must be frequently monitored. Staying under suspended loads must be strictly avoided. In all cases, it is necessary to maintain a safe distance from overhead lines of any type.

15.3.1.6. *Significant torsional moments during the tightening or loosening of drilling rod/tool unions*

Special attention must be paid to the proper working condition of tightening and cutting tools. In particular, spring collets must be used with great care and only by skilled personnel. Special attention must also be paid to the proper condition and safety of the workstation at the drilling exit point.

15.3.1.7. *Communication between the control cab, the drilling rig and the pipeline side*

To eliminate the dangers created by the drilling rig at the exit point by rotating tools, it is necessary to ensure continuous radio communication. With no visual contact between the machine and the pipeline sides, the use of a receiver-transmitter headset as well as walkie-talkies is recommended. In all cases, it is necessary for the operator of the drilling machine and the person in charge of the pipeline side to co-ordinate themselves before starting the work.

15.3.1.8. *Work under thoroughfares*

The risks are not limited to horizontal drilling except for the operator who "follows" the pilot hole with the receiver. Engrossed in using the receiver, he may not always be aware of his safety. It may be necessary to protect his route and/or stop the traffic temporarily while crossing the road.

15.3.1.9. *Risks of aggressions on underground structures*

The risk of sometimes coming into contact with high voltage electrical transmission lines requires a rigorous use of the drilling machine and insulation of the personnel. In the case of a damaged gas pipeline, all ignition sources must be removed: the machine and all the equipment near the leak must be stopped, as well as all electrical devices (including mobile phones). When gas pipelines are present in the drilling area, the site manager must provide the co-ordinates of the concerned gas company.

15.3.2. *Security of machines*

The machines used for carrying out horizontal drilling projects must meet the European recommendations as well as the national regulations applicable to them. A compliance statement as well as the issuing of the CE acronym linked to it must be provided by the manufacturers of different machines. Independent monitoring of the machines is possible. It will be done by the competent national authorities as far as safety is concerned. This is linked to the qualification and for *European Tested*. For protection against electrical accidents, the machines operating on electricity must be properly earthed before use. The maintenance of complex hydraulic systems of horizontal drilling machines must be done with care. The watertightness of these systems must be constantly monitored.

15.3.3. *Security of drilling tools*

For drilling rods, tools, equipment, joints, reamers and links to be inserted inside the drilling, it will be necessary to undertake a safety check by a qualified body in MQ or by a qualified or state recognized monitoring institution (independent monitoring). This monitoring must show that the drilling rods and the tools used inside the drilling are made of suitable material. It must also show that the maximum stresses to tension, compression, torsion and internal pressure incurred by the drilling machine used do not exceed 0.8 times the elastic limit of the material (S = 1.25) according to the DIN, API bases and DS standard. For tools having rotating parts such as downhole mud motors, tricone bits or universal joints, one must ensure before every use that a detailed and documented inspection is carried out. This will guarantee that they can be used safely and without any restrictions (internal inspection).

15.3.4. *Protection of the environment*

Generally the horizontal drilling technique is more environmentally-friendly than traditional open trench installation techniques.

If sensitive soil is encountered at the site, covering boards for the passage of earthmovers must protect it. For obtaining cuttings required for different excavations, it is necessary to ensure that the different layers of topsoil remain unmixed with the deeper layers of the soil.

Great care must be taken to avoid contaminating the soil with oils and lubricants (use of salvage tanks, tarpaulins, etc.).

Particular attention must be paid to storage of bore fluids. A sufficient storage capacity must be available at any moment. In any case, all uncontrolled leaks of bore fluids must be avoided.

The removal of residual bore fluids at the end of the work and the extracted cuttings must be organized before starting the project. It is necessary to restore the work area followed by an acceptance in the presence of different landowners.

Appendix 1

Glossary of Symbols Used

Roman symbols

b	Width of the ground affected by Terzaghi's silo theory
c	Coherence of the soil
c_u	Undrained cohesion of the soil
c_{ur}	Undrained cohesion of the reworked clay
D_e	Diameter of the excavation
D_{ext}	Outer diameter of pipes
DP	Additional thrust P in the jacks during start-up
EH	Horizontal deviation
E_M	Pressiometric module
EV	Vertical deviation
E_s	Young's modulus of the soil
f	Unit soil-pipe dynamic friction
f*	Unit soil-pipe static friction
f_{app}	Apparent average unit friction
f_{conv}	Conventional average unit friction
f_{lub}	Unit friction in the presence of lubrication
f_{sup}	Additional friction caused by stoppages in jacking

f_{stat}	Unit static friction
F	Frictional force on the entire jacked pipeline
H	Height of the overburden above the pipes
IH	Azimuth (difference in horizontal angle between the theoretical trajectory and the machine body)
I_L	Flow index
IP	Plasticity index
IV	Vertical tilt (difference of vertical angle between the theoretical trajectory and the machine body)
k	Reduction coefficient of the vertical stress caused by arching
k_M	Marston's coefficient in narrow trenches
K	At rest earth pressure ratio = σ_h / σ_v
K_a	Active pressure coefficient
K_p	Passive pressure coefficient
L	Length of jacked pipes
M_d	Impact factor for rail overloads
n	Normal stress of the soil being applied on a pipe section of unit length
N	Total normal stress being applied on all the pipes
pl	Limit pressiometric pressure
P	Thrust developed by the jacks in the thrust shafts
P_{total}	Total thrust developed by the jacks in the thrust shafts
q_o	Effect of an overload q_s at the surface on the pipes
q_s	Overload at the surface
R_p	Thrust at the head of the boring machine
T	Jacking stoppage time
T_c	Stability coefficient for coherent soil
T_γ	Stability coefficient for frictional soil
T_s	Stability coefficient caused by an overload
W	Dead load of pipes
w	Water content
w_L	Liquid limit
w_P	Plastic limit

Greek symbols

α	Angle between successive pipes, or between the head and the body of the boring machine ("angulation")
β	Characteristic coefficient of the soil-pipe adherence
δ	Soil-structure frictional angle
Δ_h	Horizontal convergence in the excavation diameter
Δ_v	Vertical convergence in the excavation diameter
Δ_p	Increase in diameter caused by a pressure p inside the overcut
γ	Specific weight of the soil
φ	Internal friction angle of the soil
μ	Friction coefficient
$\mathbf{v}_s$	Poisson's ratio of the soil
σ	Normal stress on a facet
σ_0	Average stress in the pipes
σ_{max}	Maximum stress in the pipes
σ_c	Unconfined compression strength
σ_{EV}	Vertical stress on the horizontal plane at the pipe wrench
σ_h	Normal stress on a vertical facet
σ_T	Confining pressure required for the stability of the excavation
σ_v	Normal stress on a horizontal facet
τ	Shear stress on a facet

Symbols specific to Chapter 8

CD	Direct cost of works
CG	Overall cost of works
CS	Social cost of works
α_{CS}	Multiplier for calculating the social cost
TET	Trench technique
TST	Trenchless technique

Appendix 2

Glossary of Horizontal Drilling

API: American Petroleum Institute. It defines and governs the majority of the standards relating to the techniques of drilling.

Attapulgite clay: clay which increases the viscosity of mud in salt medium. Used in the place of bentonite.

Backreamer; hole opener, expander: tool used to enlarge the hole carried out during the pilot hole.

Backreaming: backreaming of the diameter of pilot hole. This operation can be carried out in several stages, according to the dimension of the final hole.

Bent sub: inserted between the stem and the tool, it enables the tool to operate at an angle (of a few degrees).

Bentonite: sodium colloidal clay of Montmorillonite type used in drilling mud. Concentration: from 30 to 100 kg per m^3.

Biodegradability: capacity of the components of mud to be broken up under the action of biological agents present in nature.

Bore, diameter: diameter of drilling or reaming.

Box: female threading of a drilling rod.

Breakway connector: inserted in between the reamer and the drilling head, it makes it possible to limit the force exerted on the sleeve (or the pipeline). In the event of the limit being exceeded there is a breakaway.

Cake: lining or film of variable thickness deposited by the mud on the walls of the tunnel. Consolidates and waterproofs the drilling.

Calibration: tuning of the receiver of a system of guidance in order to get an exact and precise measurement of the depth.

Caustic soda: can be used to increase the pH of acidic mud.

Caving: collapse of the borehole, often due to an insufficient mud pressure to compensate for the hydrostatic pressure of the ground.

Cavitation: phenomenon occurring when one pump is not fed under a sufficient pressure. It is characterized by a noise of cracking and can seriously damage the pump.

Clay: particles of size lower than 2 microns (decimal system of Atterberg).

Claystone: strengthened clay.

Cleaning: mud's capacity to clean the tool and the hole from the spoil resulting from drilling.

CMC, carboxy methylcellulose: organic colloid used as reducer/reducing agent of filtrate.

Coarse soils: sands and gravels.

Cobbles: elements of size ranging between 20 and 200 mm (Atterberg decimal system).

Compactor bell: cone shaped reamer for compressible grounds.

Crossing: passage of special points such as rivers, channels, railways, motorways, etc.

Cuttings: materials resulting from drilling. They are transported and evacuated away from drilling by the mud.

Desander: unit making it possible to recycle the drilling mud by eliminating the largest part of sand.

Differential pressures: in a mud engine, it is the difference between the pressure under operation and the pressure during tuning of the turbine. The value recommended by the manufacturer of the engine should be adhered to.

Directional boring systems, directional drilling systems: technique enabling the laying of pipelines drains without trenching.

Drag bit: cutting tool equipped with steel, for the mud engines (tender rocks).

Drill unit, drill carriage: element of the drilling machine on which is located the active part of the drilling machine. Thermal engine and hydraulic unit can form part (autonomous drilling machines) of it.

Duckbill, Drill bit: blade fitted at the tip of the drilling head. Some are equipped with points or teeth for the hard grounds.

Entry pit: pit at the starting of a directional drilling making it possible to channel and recover mud at the time of pilot drilling.

Exit pit: pit at the end of the directional drilling making it possible to recover and channel mud at the time of backreaming.

Filter loss; Fluid loss: percentage of water, which passes through the "cake" and leaks out from the borehole into the ground. The formation plays the role of filter and the solid particles of the mud collected on the walls of the tunnel will form the cake. It is measured with a "filter press".

Filter press: apparatus allowing to measure the filter capacity of a mud and thus evaluate its aptitude to form a "cake" of good quality. The quantity of water which passes through a filter under a given pressure is measured.

Filtration: mud's capacity to form an impermeable film on the walls of the tunnel. Measured with a filter press.

Fluid loss: mud does not come out any more, either on one side or on the other side of the borehole, because of the porosity of the ground or the presence of fractures. It is necessary to modify mud or to add plugging materials.

Fluid loss agent: when added to mud, this substance decreases the value of the filtrate (reduces losses of liquid during the formation).

Fluted reamer: compacting styled reamer equipped with points or edges; general-purpose.

Formation: area of ground having homogeneous physical and chemical characteristics.

Frac out: when the pressure of the drilling mud becomes too strong, fractures may appear on the surface, which then will allow mud to leak.

Fuse head: element placed between the backreamer and the pipe-puller which, by rupture, avoids exceeding the limit of the traction value on the polyethylene pipelines.

Gel strength: characteristic feature of mud to freeze when it stops being agitated. It will determine its aptitude to maintain solid particles ("cuttings") in suspension for transporting them.

Ghost: parasite/interfering signals coming from external electromagnetic waves received by the system of guidance.

Gravel: elements of size ranging between 2 and 20 mm (Atterberg decimal system).

Ground stake: copper stem screwed in the ground making it possible to hammer the drilling machine in the event of electric shock.

Hardness: characteristics measured by the resistance to simple compression (Rc) of a sample of rock. This test, carried out in a laboratory, gives a reading expressed in mega Pascals (MPa). 1 MPa = 10 bars = 140 PSI.

Hydra-lock: under certain conditions (insufficient flow, mud loss or too fast boring), the borehole can be closed again on the pipelines, which then causes a rise in pressure of upstream mud and can lead to a blocking of pulling.

Jetting: technique of drilling in which the pressure of liquid breaks up the ground. In that case, this pressure can reach 400 bars.

Marl: sedimentary ground made up of a mixture of clay and limestone.

Marsh funnel: funnel equipped with sieves with 20 mesh and a calibrated opening making it possible to measure viscosity of mud.

Milled tooth bit: drilling tool equipped with serrated rollers with steel teeth (for not very abrasive hard rocks).

Montmorillonite: a variety of clay having physicochemical properties and the rheological qualities particularly suitable for drilling.

Mud motor: special drilling head meant for rock. Consisting of a hydraulic turbine driven by the drilling mud and functioning like an Archimedes' screw. This turbine drives in its turn a tool: tri-blade or tri-cone.

Mud pressure: this is the mud pressure in the borehole. It must balance the hydrostatic pressure exerted by water of the ground. If it is too weak, there is a risk of caving; if it is too strong, there is a risk of frac out.

Mud pump: high-pressure pump located on the drilling rig or at the mud unit. It is important that it is always charging up in order to avoid cavitation.

Mud return: removal of the earth spoil away from the workplace by the mud.

Mud, slurry: liquid used in directional drilling. Constituted mainly of bentonite and/or polymers and additives mixed with water. NOTE: the surface-active ones such as the detergents accelerate the ageing of polyethylene.

Mud system: unit used to manufacture the drilling mud.

Mud weight, mud density: mass of one unit of mud per unit of volume (kg/m^3, g/cm^3, etc.). Its value will influence the pressure of mud, and therefore the stability of the tunnel.

Nozzles: openings gauged on the drilling head and reamer for mud injection. If they are big, they increase the flow; if they are small, they increased the pressure of the mud flow (jetting).

pH: unit of measurement of acidity (concentration in H + ions) of a liquid. A liquid is acidic from 0 to 7, neutral with 7 and alkaline from 7 to 14. This pH should generally be maintained between 8.5 and 9.5 (slightly alkaline).

pH-meter: apparatus which enables a fast measurement of the pH of a liquid. More practical for use than the colored strips.

Pilot hole: first phase of a directional drilling. This determines the trajectory of drilling.

Pin: male threading of a drilling rod.

Pipe-puller: obturator equipped with a ring to draw the pipelines.

Pitch: angle of the drilling head with respect to the horizontal. It is displayed in degrees or in % on the receiver of the guiding system.

Plugging material: additive with mud made up of particles whose dimensional specifications make it possible to clog the ground and avoid losses of liquid (fractures, porous grounds, etc.).

Polymer: long chain molecule generated by the repetition of a small structure called “motif”. Many polymers are used in the drilling mud. They reduce the swelling of clays and improve the quality of the cake.

Pre-reaming: for the large diameters or for difficult grounds, several successive backreamings are necessary, with more and more large reamers.

Pullback force: force exerted by the drilling unit on the drill string at the time of backreaming. Expressed in tons.

Rack angle: starting angle of the drilling with respect to the horizontal. It can be measured directly with the guidance system.

Rack, feed frame: beam on which the drilling unit slides.

Recycling: some sites require large quantities of mud (drilling in the rock). For economic and ecological reasons, it can be recycled.

Remote, drilling parameter display: mounted on the drilling machine, it displays the parameters of drilling of the guidance receiver system.

Removal: this is one of the principal roles of the drilling mud. In the case of rock, which is not compressible, large quantities of mud will be necessary to evacuate the solids (“cuttings”), especially at the time of backreaming.

Rheology: study of the deformation and flow of the matter: viscosity, freezing, etc.

Rock drilling: possible for rocks up to 300 MPa. Requires the use of specific tools (rock engine, rock reamers, probes with cable, etc.).

Rod, drill pipe, stick: threaded tubular element enabling the progression of drilling and the transport of mud.

Rod recycler: the part which makes it possible to draw a second drill string behind a reamer in the case of multiple backreaming.

Rod wiper: neoprene disc placed at the exit of the drill string which enables to clean them from drilling mud during backreaming.

Roll: position of the drilling head with respect to its axis, expressed from midnight to 12 noon. Only this position will make it possible to deviate the drilling in the desired direction.

ROP (rate of penetration): speed of progression of the pilot hole.

Sand: particles of size ranging between 20 microns and 2 mm (Atterberg decimal system).

Sand content: contents of particles of more than 74 microns (sieve of 200 mesh) in the mud. Measured with a sand content set.

Sand content set: sieves with 200 mesh and a graduated test-tube used to measure the sand content of mud (particles > 74 microns). It is desirable that this content does not exceed 2%.

Shale: Metamorphic rock.

Silts: particles of size ranging between 2 and 20 microns (Atterberg decimal system).

Slurry level: it is desirable that this level is maintained at a minimum of 2 meters at least above the level of groundwater in order to counterbalance the hydrostatic pressure of the ground.

Soda ash: used to increase the pH of mud (reduces the acidity) and to decrease its hardness (magnesium and calcium salts).

Soil remediation: use of the directional drilling for the laying of drains in order to drain and clean the contaminated grounds (hydrocarbons, etc.).

Sonde housing: probe housing of the steering system; this housing can be in the head itself or a casing located between the head and the first rod.

Sonde, probe, transmitter: transmitting element of the detection and guidance system.

Specific gravity: weight of a liquid with respect to water. Influences the hydrostatic pressure of mud and controls the ejection of water from the ground. Conversely, if it is too high, it risks breaking the tunnel and leading to circulation losses.

Spindle torque: couple exerted on the train of stems by the drilling unit. More than the pullback force, it concerns of the major characteristic of a drilling machine. Expressed in Newtons.meters (1 N.m = 9.81 m.kg).

Starchy: products made up of carbon hydrates, prepared from the starch of some plants. Can be used as filtrate reducing agent.

Steering system: a structure which enables the detection and the guidance of the drilling head, consisting of a transmitter (in the drilling head), a receiver and generally a drilling parameter display on operator's desk of the drilling machine.

Sticky clays: these clays can block the drill string in the ground. The problem is solved with anti-sticky additives.

Strike alert: system of visual and sound alert activated as and when the head or the drill string touches an object under tension.

Stripes: longitudinal notches created on a polyethylene pipeline at the time of pulling. A maximum depth of 10% of the thickness of the tube is permitted for gas pipelines.

Stripes measuring unit: makes it possible to measure the depth of the stripes on the polyethylene pipelines.

Sub: threaded mechanical part (male or female) enabling the connection of drilling elements (stems, casings, reamers, etc.).

Swelling: increase in volume of clays by absorbing water.

Swelling clays: clays capable of swelling in volume by absorption of water (coefficient of swelling $C_g > 0.04$).

Swivel: the part which enables the rotation; used to draw the pipeline without making it rotate.

TCI bit: drilling tool equipped with serrated rollers with carbide (for hard, very hard and abrasive rocks).

Thrust force: force exerted by the drilling unit on the drill string at the time of pilot hole. Expressed in tons.

Transfer pump: low pressure pump which drives back the mud from the tank towards the high pressure pump.

Tricone bit: cutting tool mounted mainly on the mud engines for drilling in the rock. It is generally equipped with three serrated rollers assembled on bearings, watertight or not.

Venturi: system of aspiration of bentonite by depression in the mixing unit.

Viscosity: resistance to fluid circulation; measured with a viscometer (time of flow through a gauged opening in seconds: 26 seconds for pure water).

Water hardness: calcium and magnesium content of water. Alters the properties of bentonite. This can be solved by adding calcium carbonate.

Weight on bit (WOB): pressure (tons or kN) exerted by the drill string on the drill bit. The manufacturers the tri-cone tools provide a measurement limit, in order to avoid damaging of the bearings.

Wing cutter: reamer provided with teeth to cut out the ground; suitable for clayey grounds.

Wireline steering system: guidance system where the information goes up to the receiver through a cable introduced into the stems. Its range is no longer limited, nor is its autonomy (no batteries, the cable ensuring also the power supply).

Bibliography

General references

AFNOR (1992), Norme NF P 11-300, "Classification des matériaux utilisables dans la construction des remblais et des couches de forme des infrastructures routières", September 1992.

AFNOR (2000), Norme sur la mise en oeuvre sans tranchée et essai des branchements et collecteurs d'assainissement, Classement P16-126, 2000.

AFTES (1976), Texte de réflexions sur les méthodes usuelles de calcul du revêtement des souterrains, *Tunnels et Ouvrages Souterrains*, Paris, no. 14, March-April 1976.

AFTES (1982), Texte de réflexions sur les méthodes usuelles de calcul du revêtement des souterrains, *Tunnels et Ouvrages Souterrains*, Paris, July 1982.

AFTES (2002), "Recommandations relatives à la boue de forage à usage des boucliers à pression de boue", *Tunnels et Ouvrages Souterrains*, no. 171, May-June 2002.

AIT-AISSA D. (1997), Impact des chantiers urbains: les techniques nouvelles qui réduisent la gêne coûtent-elles plus cher? L'exemple des technologies sans ouverture de tranchée, thesis in Urban Engineering, Ecole Nationale des Ponts et Chaussées, University of Technology, Compiègne, France, 1997.

ANGOT C. (1991), Prise en compte du coût social pour la promotion de techniques innovantes de pose de conduites, final essay for the Professionnal Master's Degree (DESS) in Technology and Innovation Management, University of Paris IX Dauphine, Compagnie Générale des Eaux, 1991.

ANHEUSER L. *et al.* (1994), *Maschineller Tunnelbau im Schidvortrieb*, Ernst & Sohn, Berlin, 1994.

BEAUVAIS J.M. (1978), *Coût social des transports parisiens*, Economica, Paris, 1978.

BENNET D., CORDIN E.J., ISELEY T. (1994), Auger and slurry microtunneling tests under controlled ground conditions, Transportation Research Record, 1994, no. 1431, pp. 75-82, 1994.

BENNET D., GUICE L., STAHELI K., KHAN S. (1995), Guidelines for trenchless technology, US Army Corps of Engineers, Washington, Final Report CPAR-GL-95-2, 1995.

BEROSCH M. (1994), Prise en compte du coût social pour la pose et la réhabilitation des réseaux d'eau et d'assainissement, Formation Continue ENPC, Techniques de forage sans tranchée: microtunnels et forages, 1994.

BIELECKI R., SCHREVER J. (1997), "Guidelines for selecting construction methods for earth-laid lines", *NO-DIG.97*, Genes, Italy, International Society for Trenchless Technology (ISTT), 1997.

BOYCE G.M., GRAY W.W.S. (1996), "A User's guide of pipes", *International NO-DIG.96*, ISTT, New Orleans, USA, pp. 519-532, April 1996.

COLLER P., STAHELI K., BENNETT D., POST R. (1996), "A review of jacking forces by both theoretical and empirical methods as compared with 20 years of pratical experience", *Proceeding of International NO-DIG.96*, New Orleans, USA, pp. 126-150, April 1996.

DTU 13.2 (1983), Calculs des fondations profondes soumises à charge axiale, CSTB, Paris, Chapter 11, file no. 1877, September 1983, 7th edition, April 1987.

FSTT (1994), "Objectifs et moyens des études de sol pour les projets de microtunnels", *Cahiers techniques de la FSTT*, no. 1, May 1994.

FUJITA (1990), "The legal system for road works in Japan in the case of sewer construction", *NO-DIG.90*, Osaka, Japan, International Society for Trenchless Technology (ISTT), 1990.

GUILLOUX A., LEGAZ C. (1992), "Synthesis of microtunneling projects in the Val de Marne", *Proceedings Int. Conference NO-DIG.92, No trenches in town*, Balkema, Paris, pp. 151-157, October 1992.

HASLEM (1986), "Pipe jacking forces: from practice to theory", *Infrastructure Renovation and Waste Control: International Centenary Conference*, Manchester, UK, April 1986.

JSTT WORKING GROUP NO 3 (1994), *Microtunneling jacking force*, Japan Society of Trenchless Tunneling Techniques, Japan, May 1994.

KANARI H. (1998), "Requirements for lubrificants and friction-reducing effects of slow hardening lubricants", *Proc. Of the 16th International ISTT NO-DIG.98*, Lausanne, Swiss, 1998.

KASTNER R., PELLET A.-L., OUVRY J.-F., GUILLOUX A. (1996), "In situ monitoring of microtunneling projects", *Proc. Intern. NO-DIG.96*, New Orleans, USA, pp. 171-182, April 1996.

LAURITZSEN R., SANDE O., SLOTTEN A. (1984), "Europipe landfall tunnel", *Proc. of International NO-DIG.94*, article G2, Copenhagen, Denmark, April 1994.

LENGLET F. (1992), "L'Europe doit repenser ses transports", *L'Expansion*, no. 427, Paris, pp. 34-41, 1992.

LEROUEIL S., TAVENAS F., LE BIHAN J.-P. (1983), "Propriétés caractéristiques des argiles de l'est du Canada", *Canadian Geotechnical Journal*, vol. 20, pp. 681-705, 1983.

MARCHAND L. (1984), "Un concept fécond: la consommation espace-temps", *Cahiers scientifiques de la revue des transports*, Paris, 1984.

MARSHALL M.A., MILLIGAN G.W., MAIR R.J. (1996), "Movements and stress changes in London Clay due to construction of a pipe jack", *International Symposium Geotechnical Aspects of Underground Construction in Soft Ground*, London, Balkema, Rotterdam, Holland, pp. 629-634, 15-17 April 1996.

MERMET M., CROSNIER D., MERCENNE B. (1991), "Techniques sans tranchées en France (FSTT)", *Proceedings 7th Int. Conference on Trenchless Construction for Utilities, NODIG.91*, Hamburg, Germany, October 1991.

MICHELIZZA (1991), "Entreprendre une activité de microtunnelier, aspect économique et marché", *ENPC, Microtunnels et microtunneliers*, 17-18 December 1991.

MILLIGAN G.W., MARSHALL M.A., (1995), "Ground movements due to construction of pipe jacked tunnels", *Proceeding of 11th Euro. Conf. Soil Mech. And Fnd. Eng.*, Copenhagen, Holland, vol. 3, pp. 191-200, 1995.

MILLIGAN G., MARSHALL M.A., (1998), "The influence of lubrification on jacking loads from six monitored pipe jacks", *Proc. Of the 16th International ISTT NO-DIG.98*, Lausanne, Swiss, 1998.

MILLIGAN G., NORRIS P. (1995), "Pipe Jacking Research Results and Recommendations", *Pipe Jacking Association*, London, 1995.

MILLIGAN G., NORRIS P. (1999), "Pipe soil interaction during pipe jacking", *Proc. Instn Civ. Engrs, Geotech. Engng*, no. 137, pp. 27-44, January 1999.

MONFRONT L. (1994), "Etat de l'art et recommandations pour le dimensionnement des tuyaux pour microtunnelage", *FSTT, Cahier technique no. 2*, September 1994.

NORGROVES W., O.REILLY M. (1990), "Comparisons of the costs of constructing sewers by tunnelling or trenching in urbans areas", *Proc. 5th Int. Conference on trenchless construction for utilities NO.DIG.90*, Rotterdam, Holland, pp. A6.1-A6.10, 1990.

PARKER H.W. (1999), "Geotechnical Investigations for Tunnels", *CR des Journées d'études AFTES*, Paris, pp. 49-56, October 1999.

PECK R. (1969), "Deep excavations and tunneling in soft ground", *Proc. of 7th International Conference Soil Mechanics and Foundation Engineering*, Mexico City, Balkema, Rotterdam, Holland, state of the art volume, pp. 225-290, 1969.

PELLET A.-L. (1997), Analyse des interactions sol-canalisation et sol-machine pour la pose de conduites par microtunnelage, thesis in Civil Engineering: INSA (Institut National des Sciences Appliquées) URGC Géotechnique, Lyon, France, December 1997.

PELLET A.-L., KASTNER R. (1999), "Etude expérimentale et analytique des efforts de frottement mis en jeu lors de la pose de conduites par microtunnelage", *Annales du bâtiment et des travaux publics*, no. 6, pp. 55-72, December 1999.

PELLET A-L., KASTNER R., AIELLO, A. (1995), "Networks installation by trenchless technology, experience of a microtunneling operation using a pneumatic conveyor", *Proc. of Underground Space and Urban Planning*, Paris, pp. 481-486, September 1995.

PELLET A.-L., KASTNER R., GUILLOUX A. (1997), "Microtunneling: comparison between friction forces predictions and *in situ* measurements", *Proc. int. NO-DIG.97, ISTT*, Taipei, Taiwan, 1997.

PHELIPOT A. (2000), Interaction sol-structure lors d'opérations de microtunnelage, thesis in Civil Engineering: INSA (Institut National des Sciences Appliquées), URGC géotechnique, Lyon, December 2000.

PHELIPOT A., KASTNER R. (1997), "Microtunneling site in clay: slurry problems, effect on existing works", *Proc. int. NO-DIG.98, ISTT*, Lausanne, Swiss, 1997.

PIPE JACKING ASSOCIATION (PJA) (1995), *Guide to best practice for the installation of pipe jacksand microtunnels*, London, 1995.

ROBINSON A., THOMSON J. (1997), "Analysing the real cost of utility installation", *NODIG.97*, Genes, Italy, International Society for Trenchless Technology (ISTT), 1997.

ROGERS C.D.F., O.REILLY MP. (1991), "Ground movements associated with pipe installation and tuneling", *Proc. of 10th Euro. Conf. Soil. Mech. And Fnd. Eng.*, Florence, vol. 2, Balkema, Rotterdam, Holland, pp. 907-910, May 1991.

SCHLOSSER F., LECA E. (1992), "Microtunnels: construction d'ouvrages", *Congrès International NO-DIG.92, No trenches in town*, Balkema, Paris, October 1992.

STAHELI K., BENETT D. (1996), "Results of controlled field test of a retrieval microtunneling system with reaming capabilities", *Proc. of International NO-DIG.96*, New Orleans, USA, pp. 184-202, April 1996.

STEIN D., MOLLERS K., BIELIECKI R. (1989), *Microtunneling: installation and renewal of non man-size supply and sewage lines by the trenchless construction method*, Ernst & Sohn, Berlin, 1989.

STERLING R.L. (1997), "Indirect costs of utility construction and repair", *NO-DIG.97*, Genes, Italy, International Society for Trenchless Technology (ISTT), 1997.

SZECHY, K. (1970), *Traité de Construction des Tunnels*, Dunod, Paris, pp. 157-702, 1970.

TERZAGHI, K. (1951), *Mécanique des sols*, Dunod, Paris, pp. 188-474, 1951.

VICKRIDGE I. (1997), "Mechanisms for controlling social costs", *NO-DIG.97*, Genes, Italy, International Society for Trenchless Technology (ISTT), 1997.

VIDAL G. (1993), "La gêne des travaux en milieu urbain, étude exploratoire", Association des Ingénieurs des Villes de France (AIVF) et Institut pour la Recherche appliquée et l'EXpérimentation en génie civil (IREX).

Reports on the French National Research Project "Microtunnels"

Technical Reports

FSTT RT 1. Quebaud S., Henry J.P., Aubry C. (1995). Présentation de la base de données microdata. LML/Géodesign.

FSTT RT 2. Monfront L., Kastner R., Cotin (1995). Exploitation des mesures réalisées sur les tuyaux. BONNA/INSA/BRGM.

FSTT RT 3. Ouvry J.F. (1994). Chantier expérimental de La Vallée aux Loups. ANTEA.

FSTT RT 4. Dessauvage P., Lemee M., Quebaud S. (1996). Apport des additifs dans l'amélioration du percement par microtunneliers. LML/EUDIL.

FSTT RT 5. Delorme P. (1994). Manuel d'utilisation de la base de données "forages dirigés". GDF.

FSTT RT 6. Delorme P., Homand F. (1996). Instrumentation d'un chantier de forages dirigés à Paray Vieille Poste. GDF/Geomechanic Laboratory, Nancy.

FSTT RT 7. Mered R., Jacques J. (1996). Saisie de données et tests de la base microdata. FSTT.

FSTT RT 8. Quebaud S., Renaud V., Shao J.F. (1996). Synthèse des fiches performances/terrain : base de données microdata. LML.

FSTT RT 9. Ouvry JF., Gaillard C.(1996). Suivi léger des chantiers de microtunneliers. ANTEA/ESIP.

FSTT RT 11. Guilloux A., Lac C. (1996). Microtunnel RN 303 à Champigny. Synthèse générale du chantier. TERRASOL.

FSTT RT 12. Chazelas, (1998). Création d'un site test pour les méthodes de géophysique appliquées aux travaux sans tranchée. LCPC.

FSTT RT 13 et 14. Delorme P. (1996). Analyse et amélioration de la technique du forage dirigé pour la pose de canalisation en polyéthylène. Thesis, University of Lille, in collaboration with GDF.

FSTT RT 15. Monfront L. (1996). Comportement des tuyaux à la flexion essais en laboratoire. BONNA.

FSTT RT 16. Quebaud S., Sibai M., Shao J.F. (1996). Campagne d'essais relative à l'apport des produits moussants sur des sols grenus. Géodesign/LML.

FSTT RT 17. Quebaud S., Des Ligneris N. (1996). Utilisation de la mousse sur un chantier de microtunnel. LML EUDIL, Fougerolle-Ballot.

FSTT RT 18. Quebaud S., Sibai M., Shao J.F. (1996). Comportement du mélange argiles-mousse. LML EUDIL.

FSTT RT 19. De Sloovere P. (1999). Reconnaissance/Sismique transparence sur le site test du Bourget. ME2i.

FSTT RT 20. Ouvry J.F., Antoinet E., Nava S., Kastner R., Guilloux A. (1998). Suivi léger des chantiers de microtunneliers; synthèse de 12 chantiers. ANTEA, INSA, Terrasol.

FSTT RT 21. Bousquet-Jacq P., Guilloux A., Butterworth J., Foillard R. (1998). Reconnaissance par forages dirigés d'un tronçon de microtunnel: étude de faisabilité, suivi, diagraphies différées et synthèse. Terrasol, GEOSCAN, Le forage directionnel.

FSTT RT 22. Phelipot A., Kastner R., Bourdeau Y. (1999). Interprétation et frottement d'essais de traction des canalisations en polyéthylène. INSA, Lyon.

FSTT RT 23. Lagabrielle R. (2002). Tomographies électromagnétiques et sismiques entre forages réalisés au LRPC du Bourget Août 1998. LCPC.

FSTT RT 24. Phelipot A. (2001). Etude de l'interaction sol-canalisation sur sables. INSA, Lyon.

FSTT RT 25. Kastner R. (2001). Les distances de sécurité lors d'opérations de forages dirigés. INSA.

FSTT RT 26. Kastner R. (2002). Etude d'impact du microtunnelage sur l'environnement. INSA, Lyon.

FSTT RT 27. Pantet A., Ouvry JF., Didier G., Merlet N. (2000). Document technique et réglementaire. Traitement et recyclage des boues. ESIP/ANETA/INSA, Lyon.

FSTT RT 28. Pantet A. (2001). Caractérisation physico-chimique et rhéologique de boues bentonitiques. ESIP Poitiers.

FSTT RT 29. Wilkinson R. (2001). Le forage dirigé dans les galets. WISE.

FSTT RT 30. Pantet A., Guérin M., Monnet P. (2002). Synthèse sur les liquides diffusés en TST. ESIP/Calypso/UMR-LEA.

FSTT RT 31. Bultel, (2002). Les aspects socio-économiques (coûts sociaux, aspects contractuels) induits par les Travaux sans Tranchée. Scetauroute.

Final Reports

FSTT RS 1. Pellet A.L., Kastner R. (1994). Chantier expérimental de la Vallée aux Loups. BONNA, INSA, Lyon Valentin-Eternit.

FSTT RS 2. Monfront l. (1994). Méthode de calcul. BONNA.

FSTT RS 3. Dubernay C, Morel E. (1994). Les techniques sans tranchée et le broyage dans les terrains hétérogènes. LML EUDIL.

FSTT RS 4. Pannetier C., Homand F. (1994). Caractérisation de l'abrasivité des terrains. Geomechanic Laboratory, Nancy.

FSTT RS 5. Reanud-Casbonne F., Homand F. (1994). Le collage des argiles. Geomechanic Laboratory, Nancy.

FSTT RS 6. Delorme PH., Homand F., Belem T. (1995). Etude préliminaire de l'interface sol/polyéthylène pour la pose de canalisations en polyéthylène par forages dirigés. GDF/ Geomechanic Laboratory, Nancy.

FSTT RS 7. Panissot C., Tabbagh A. (1995). Etude d'un système multipole électrostatique (volte n° 1). CNRS, University Paris VI.

FSTT RS 8. Quebaud S. (1996). Etude expérimentale de l'interaction sol-polyéthylène. Thesis, University of Lille, in collaboration with GDF.

FSTT RS 9. Pellet A.-L., Kastner R., Guilloux A. (1996). Suivi expérimental du tronçon n°4 du chantier de microtunnelage de Champigny. INSA, Lyon, Terrasol.

FSTT RS 10. Foillard R., George B. (1995). Expérimentation radar sur le site-test tomo-collecteur du LREP. GEOSCAN.

FSTT RS 11. Pellet A.L., Kastner R. (1996). Chantier de microtunnelage de Neuilly: Suivi expérimental. INSA, Lyon.

FSTT RS 12. Quebaud S. (1996). Contribution à l'étude du percement de galeries par boucliers à pression de terre: amélioration du creusement par l'utilisation de produits moussants. Thesis, University of Lille.

FSTT RS 13. Erling J.C., Cailleux B. (1994). Site-test du Bourget: auscultation géophysique en forage; méthode du cylindre électrique corrélée avec les méthodes géophysiques de surface. EDG, LREP.

FSTT RS 14. Pellet A.-L., Kastner R. (1996). Chantier de Montmorency: suivi expérimental de deux tronçons. INSA, Lyon.

FSTT RS 15. Caizergues C., Delorme P., Jolivet A., Panissod C., Tabbagh A. (1996). Reconnaissances géophysiques pour travaux de forages dirigés. GDF, CNRS.

FSTT RS 16. Franquet C., Guilloux A., Leca E., Mermet M. (1997). Application du forage dirigé et du microtunnelier à la reconnaissance des ouvrages souterrains. Terrasol, LCPC, Scetauroute.

FSTT RS 17. Grangean G. (1998). Reconnaissance structurale et détection d'obstacles enfouis par radar géologique.

FSTT RS 18. Panissod C., Tabbagh A. (1997). Reconnaissance structurale et détection d'obstacles sur le site-test de Nantes par traînés et sondages électrostatiques. CNRS.

FSTT RS 19. Mechler C., Camerlynck C. (1997). Interprétation des sections radar par représentation cartographique des cartes de temps.

FSTT RS 20. Grenier G., Leonardi S., Guilloux A., Mermet J.P. (1998). Application des microtunnels au présoutènement de grandes excavations (cas de la station Monastiraki du métro d'Athènes), Canpenon-Bernard, Terrasol, Scetauroute.

FSTT RS 21. Panissod C., Tabbagh A. (1998). Logiciels de modélisation 3D pour des mesures électrostatiques à grands écartement. CNRS.

FSTT RS 22. Phelipot A., Kastner R. (2001). Suivi expérimental du chantier de microtunnelage de Barr en Alsace.

FSTT RS 23. Pellet A.L., Kastner R. (2001). Suivi expérimental du chantier de microtunnelage de Bouliac (Bordeaux). INSA, Lyon.

FSTT RS 24. Phelipot A., Kastner R. (2000). Suivi expérimental du chantier de microtunnelage à Limoges. INSA, Lyon.

FSTT RS 25. Beaucour A.L., Kastner R. (2001). Suivi expérimental de chantiers de microtunnelage. INSA, Lyon.

FSTT RS 26. Phelipot A Kastner R. (2000). Suivi expérimental du chantier de microtunnelage à Genève. INSA, Lyon.

Index

Microtunneling

Horizontal drilling